THE NAKED NEANDERTHAL

THE NAKED
NEANDERTHAL

A New Understanding of the
Human Creature

LUDOVIC SLIMAK

PEGASUS BOOKS
NEW YORK LONDON

THE NAKED NEANDERTHAL

Pegasus Books, Ltd.
148 West 37th Street, 13th Floor
New York, NY 10018

ISBN: 978-1-63936-616-3

10 9 8 7 6 5 4 3 2

Printed in the United States of America
Distributed by Simon & Schuster
www.pegasusbooks.com

Contents

Neanderthal in Our Heart and Soul

Another Intelligence

On 19 October 2017, the Pan-STARRS1 telescope at the University of Hawaii detected a cake-shaped object a few hundred metres across moving at high speed away from our sun. Telescopes on all continents were immediately trained on the fireball. They had to be quick about it – the strange object was travelling at more than 87 kilometres a second. This strange cake-like thing was nothing less than the first interstellar object ever observed in our solar system. It was swiftly named Oumuamua, literally 'first distant messenger' in the Hawaiian language. Apart from its surprising shape, it revealed anomalies never before observed on similar objects such as meteorites or asteroids: it was highly but intermittently reflective, had a weak thermal emission and accelerated surprisingly quickly once it had passed close to the sun. Abraham Loeb, the director of the Institute for Theory and Computation at Harvard University, suggested in *Astrophysical Journal Letters* that 'Oumuamua had been nothing less than humanity's first contact with an artifact of extraterrestrial intelligence'. This hypothesis was hotly debated, but it was the view of rigorous scientists in one of the top institutions in the world, and it ignited media attention around the globe.

The very hypothesis, the tiniest possibility, of an interstellar visitor captures everyone's imagination. The reason that it exerts such fascination is that it involves a form of intelligence external to humanity. A complete intelligence, fully conscious of itself and the immense complexity of material reality. But an intelligence that is not ours.

This interstellar perspective, this suggestion of distant intelligences, reminds us that we humans are alone, orphans, the only living conscious beings capable of analysing the mysteries of the universe that surrounds us. There are countless other forms of animal intelligence, but no consciousness with which we can exchange ideas, compare ourselves, or have a conversation.

These distant intelligences outside of us perhaps do exist in the immensity of space – the ultimate enigma. And yet we know for certain that they have existed in a time which appears distant to us but in fact is extremely close.

The real enigma is that these intelligences from the past became progressively extinct over the course of millennia; there was a tipping point in the history of humanity, the last moment when a consciousness external to humanity as we conceive it existed, encountered us, rubbed shoulders with us. This lost otherness still haunts us in our hopes and fears of artificial intelligence, the instrumentalized rebirth of a consciousness that does not belong to us.

In our imaginations we create our own unsettling fantasies and form our own images of this ambiguous, disappeared humanity. However, this consciousness external to us, this extinct intelligence, has so far been defined purely on the narrow basis of human intelligence such as we understand it in the present.

The Neanderthal is one of these distant intelligences, among many others, like the Denisovan or the odd tiny humans discovered on the island of Flores. And of all these extinct intelligences, it is probably the most fascinating. The Neanderthals are the original 'last savages' that are discovered afresh by each generation, from Herodotus to Columbus, from Jean-Jacques Rousseau to the island of Bougainville, from Ishi* to the 'gentle Tasaday', that much fantasized Stone Age tribe which in 1971 played the enviable role of the 'last cavemen' in the western imagination. Each generation discovers its own 'last savages', and we have ours, of course. They pop up from time to time in the media, the recurring last gasp of an immense prehistory. There has been a continuous line of them over the millennia, all fulfilling our dreams of lost worlds, from the Yeti and the Barmanou to the shores of Jules Verne's mysterious geography.

The last Neanderthals take us to an unknown universe where other consciousnesses haunt abandoned wastelands. Despite our colonization of every corner of the natural world, despite our conquest of every inch of our planet, despite all our efforts to destroy natural species, these intelligences refuse to disappear. They continue to haunt our representations of the real – on its fringes, at the world's extremities, on islets, in valleys, continents, places of refuge, scrubland, liminal spaces, rooted in an uncertain geography, vacillating between Mu and the intangible universe of Hugo Pratt's *The Celts*.

Their distant testimonies tell us that the Neanderthals were never other versions of us – not brothers, not cousins – when it comes to mental structures, but an utterly different humanity. To approach them is to encounter a fundamentally divergent consciousness.

---

* 'Ishi' is the name of the last surviving member of the Yahi Indians of California, an unknown society that died out along with Ishi at the start of the twentieth century.

Facing the Creature

For the last thirty years I have spent a large part of my time scrabbling through earth on the floors of caves. Not any old caves, and not any old earth, but soil still haunted by the presence of Neanderthals. Thirty years hunting the creatures, squeezing into the cracks and fissures where they lived, ate, slept, encountered other humans, their own and other kinds. Or sometimes where they died. And yet, after thirty years of running my hands through that earth, the mud of those caves, I am no nearer forming a clear picture of the Neanderthals. I extracted material and analysed it, procrastinated, often came to conclusions, before realizing that my theories didn't hang together. Especially at the beginning, since when you see the creature from afar, you have a deceptive sense that it is very obvious and very easy to understand.

The archaeologist, like the anthropologist, should force themselves to see both from near and from far, to paraphrase Claude Lévi-Strauss. But can we approach the Neanderthals from an anthropological point of view? Rousseau wondered whether the great apes were human, contrary to those who denied the humanity of 'savages', even though they were in fact *Homo sapiens* of a different culture. The borders of humanity have always been uncertain and indistinct, and many societies consider animals to be on a level with humans, shifting the centre of gravity, resituating humans as merely part of a whole. A whole that is infinitely more subtle than the one we can perceive by means of our social constructions alone, which artificially remove and isolate humans from their environment. What is the place of the Neanderthal in this labyrinth? Human or creature – which will take precedence in our unconscious?

Numerous textbooks describe the salient features of this extinct humanity: the lack of chin, the receding brow, the thick brow ridge above the eyes, the brain capacity, larger than ours. Small, stocky, robust, good at making things with their hands, they shared a common ancestor with us, more than 400 millennia ago. These same textbooks will point out the remarkable musculature, or the mechanics of the fingers, which create a grip that is a little different from ours. They will talk about the vast territories that the Neanderthals covered, from the shores of the Atlantic to the approaches to the Altay, the vast mountains separating the west of Mongolia from the Siberian plain. And they will recount their rather sudden extinction around forty millennia ago. The closing pages of these textbooks offer often allusive conclusions about what they really were, since of course neither the shape of our skulls nor the curve of our femurs nor the position of our thumbs has ever defined what it is that makes us human beings. And the best of these textbooks will, in their final chapters, venture beyond the material evidence of bone morphology and tentatively confront this uncertainty.

The truth is, we have not yet been able to define the inner nature of this other humanity. And such is the full extent of our uncertainty that we come face to face with the undefinable nature of humankind and the other human beings with whom, for a while, we shared our planet.

Imagine you are overlooking a vast landscape from the top of a mountain. As the land stretches away into the distance beneath you, you have the feeling that you can encompass it all within a single gaze. But this landscape, seen from on high, is just an expanse of distant reliefs, sublime, picturesque, perhaps, but which tell you nothing of the people who occupy these valleys, of the streets of these villages which to you appear no more than clumps of buildings, of how the bread in that little baker's tastes. From up there you can see for miles but you never meet another soul. And what do you know about the smells emanating from that restaurant, of the grain of the stone in the wall of that church? As you get closer, you begin to see the labyrinths of these little villages, the narrow streets that have contained the lives and hopes of a hundred generations of humans.

But the picture is still too impressionistic, and you might conclude that this immense jigsaw puzzle of 300 millennia of human history has too many missing pieces and you need to use your imagination to fill in the gaps. But that does not capture it. The creature escapes us. The Neanderthal stubbornly remains an enigma. If you think otherwise, you can only have dipped your fingers in the mud, or searched in a half-hearted way. Researchers who talk about the creature fall into two broad categories: those who are sure they know what it was and those who have questions and doubts about its true nature. The first category seems to be predominant, going by the titles in the major scientific journals, which they seem to fill. The second category is much more discreet, because those who have doubts are usually more reserved in expressing themselves. It is usually this more silent type who has mud under their fingernails and who is constantly scratching around and examining the remains left by the creature. So who the hell was this Neanderthal?

How can you talk about Neanderthals if you have not explored their stone lairs, if you have not uncovered thousands of objects that they abandoned or stashed in the corners of cliffs? To talk about the creature without ever seeing with your own eyes the spaces where it lived, without having tracked it for decades, like a hunter stalks his prey, is akin to talking in a void. As a bare minimum, you need to have directly extracted evidence from these cave archives for decades in order to say anything worthwhile about this extinct humanity. To imagine that you have anything pertinent to say about it when your only encounter is through boxes in museums is nonsense in my view. Neither flints nor the skeletal traces of prey nor even rare finds of carnal remains have any meaning when encased within four white walls. To have any hope of perceiving their true significance, you have to handle the raw material in the caves. Go searching down the same bushy paths as they did. A few months of dilettante archaeological digging will allow you to get a taste, but without the odour, and perhaps without any real understanding of what you hope to draw from it.

A visit to Neanderthal earth cannot be a casual one, and Neanderthal life cannot be lived by proxy. Archaeologists can no more hope to understand this humanity by opening drawers in a museum than ethnologists can understand a society by looking at antique feather boas behind a sheet of glass or by consulting old black and white photo albums.

And yet we are well aware that the creature will not bend itself to our wishes and desires. As a shy humanoid, it is one of the most unfathomable creatures one could attempt to confront.

A creature a bit like the creature of Frankenstein, who, in trying to create a life, created a thing, endowed with his own consciousness, which he was no longer able to master. Unfathomable, since it is hidden in the shadows of the dead, without thoughts, without any words of its own.

Forty-two millennia after its disappearance from the world of the living, researchers, experimenters and various sorcerer's apprentices are trying to give a voice to these vestiges of a humanity reduced to silence by biological extinction. To bring the creature back to life from a collection of corpses. For some people this has become something of a quest, like a search for the Holy Grail. Do we have the audacity to give a voice to a vanished form of humanity, spirits with neither glass nor alphabet? What a strange, macabre game of ventriloquism.

To give voice to this dead matter, this mute thing, we have to plunge our hands in the dust of the caves. Scratch the earth, dig out millions of flints, bones, coals. But such proofs of past existence speak to us only thanks to an alchemy between reason and imagination, in the stills of our conceptions and our representations, the real petri dishes of our theories.

And there is the creature suspended from a thread, like a pendulum swinging between facts and representations, between likeness and otherness; it is us, it is other, it is us, it is other . . .

Poor creature, a disjointed puppet imprisoned in our mental games.

So who the hell was this Neanderthal, then?

For me, the creature has become something like an old travelling companion, one of those guys you walk with but at the end of the day don't know a great deal about. I've been told so many times that it was basically the same as us. But was it really what we are? Good question.

I have the maddening feeling that instead of understanding the Neanderthal better as we pursue our common course, we are progressively fashioning it in our own image. The very idea that a creature with consciousness of itself could be fundamentally different from us revolts us. So we invent and reinvent the Neanderthal. Not that we get any clearer an image. We dress the creature up, narcissistically, like we hang clothes on a scarecrow. Having disappeared from the earth, the Neanderthal has been transformed, is still being transformed, by us into a lifeless doll. Victor Frankenstein was only an experimenter, a precursor. We are the macabre inventors, the puppet masters of these marionettes from times gone by.

The creature looks impressive, of course, even scary sometimes, with all its accessories. You will have seen it too, if you have been paying the slightest bit of attention, dolled up by our fantasies, made up as a Flintstone in suit and tie, dragging his woman by the hair, punching his metro ticket.

Let us return to those who 'know' who the Neanderthal was. A hidden but nonetheless fierce war is being waged in the scientific community. On one side, those who think the Neanderthal is another us. On the other, those who think it is an archaic form of humanity, with vastly inferior intellectual capabilities. A subhuman, a quasi-human, or any other adverb that we can place before or after 'human' that is generally unflattering, except in a Marvel comic.

It is not so much a war of ideas as a war of ideologies, in which neither camp can advance without getting more and more bogged down in the mud – and I am not talking about the mud of the caves, unfortunately. So, was the Neanderthal a human somewhere between nature and culture, or a gentleman of the caves?

Exploring the Soul of the Neanderthal

In this battle of competing perspectives, the portrait that we can draw today is either too clear, too obvious, too simplified and too neat to be taken seriously, or else remarkably confusing. As we have pieced it together from the flesh of different corpses, the creature itself has managed to escape us. Not as a real historical or scientific entity but as an egregore possessing its own life. It haunts all our imaginations, including those of researchers who are not incapable of imaginative thought. So in the last few years, in the wake of archaeological discoveries, the Neanderthal has been portrayed wearing necklaces of shells and eagle claws, playing the flute, painting cave walls, inventing technology, as an armed warrior, king of the North, a vanguard of our biological ancestors, who were still confined in their cosy Asian and African territories.

The artist Neanderthal is set against its equally powerful opposite number, the pre-human of the woods, the troll of the ancient world. A creature of stone and moss. Two anecdotes spring to mind. In 2006, when I was doing my postdoctoral research at Stanford University, a reputable professor of anthropology gave a lecture on the Neanderthals. His talk related the cognitive capacity of Neanderthals to the characteristics of their archaic anatomy. Showing a slide of a Neanderthal skull, he commented: 'I don't know about you, but if I got on a plane and saw that the pilot had a head like that, I'd get off again.' Laughter all round, a well-timed joke to catch the attention of the audience. But all jokes reveal an underlying thought, and this was not an insignificant thought. Let me put this another way. A few years later, in Russia, I had a conversation with a leading light of the Academy of Sciences who never stopped telling me, 'They are different.' I prompted my companion to expand on this concept of difference. The discussion went on into the small hours: 'Ludovic, they have no soul.'

I will never be able to thank this researcher enough for saying those words. They threw a bright light on the unspoken, unconscious assumptions which underlie great swathes of our understanding of this humanity. We understand instinctively that the two concepts are irreconcilable. That we have to choose which, of the artist-painter Neanderthal and the Neanderthal of the woods, is real and which is fake. There is no middle way between the two opposing views. Is the Neanderthal a creature of the lower depths or a genius of hidden depths?

The creature is wedged into our unconscious, and, at this stage, we have to assume that it is neither one nor the other. The Neanderthal is not a brother or a cousin. It is an object of study. The Neanderthal can't in any case be compared with anything that is familiar to us in a world where difference, alterity and classification have become more than ever taboo subjects. The creature cannot fail to be subversive. And this subversion is a challenge to our intelligence. Are we really up to facing this?

The Wolf is a Wolf to Man

In the West, as in every traditional society, anyone who transgresses taboos is violently rejected, banished from the group.

If the Neanderthal were a humanity different from our own, human without being human, it would force us to transgress the deepest taboos of our society. Should we then push the boundaries of our values or should we police our thoughts and stay aligned with our values? Should we meekly direct our gaze only to that which is most comfortable, from a social point of view, to comprehend?

A need for easy answers, a certain cynicism, groupthink – all tend to lead to relativist thinking. It doesn't matter, in the end, if the truth does not exist: it can be constructed. So let us construct it – why would we want to deal with labyrinthine truths?

This truth lies in finding a subtle definition of the intelligence of a humanoid creature that cannot be summed up as us or even as our ancestor. A human that is perhaps not subjected to the mental structures which define our very understanding of what it is to *be human*. Another intelligence, separated from us by hundreds of thousands of years of independent evolution. In this sense, the creature could be seen to be as distant from us as an extraterrestrial being. A distant intelligence of which we know nothing.

In archaeology, as in ethnography, only direct evidence has real value. Even if there are whole libraries dedicated to the topic, the only thing that is relevant is a direct confrontation with the remains of these populations. Could the witness here merge with his subject? That is likely. That is also the reason why the subject constantly tends to escape us, to slip through our fingers before we can get a real grasp. We have yet to perceive the creature in any tangible form in our minds. There are thousands of books which tell the story of our research, of our representations of the Neanderthal, the structure of its bones, the cartography of its sites, its technologies or its genetics – huge encyclopedias filled with facts. But beneath their scientific veneer there is no real thought in them, no philosophy, no overview or close-up view.

If you are interested in the shape of its pelvis or the geometry of the blocks of flint it used, the encyclopedias mentioned above will furnish you with more information than you can comfortably digest. But if you are trying to imagine, even superficially, what the world of these different humanities was like, such books will only disappoint you.

So this book is something different. We need to leave the libraries and go out in the field, track the creature across its furthest-flung territories, into its rocky dens, get as close as possible, despite the temporal distance, transgress time, a little, in order to try to understand how it became extinct.

And since the subject becomes entangled with its witnesses,

I will also tell you about my own journey as a researcher and hunter of Neanderthals. You will join me on the flanks of the polar Urals, where I encountered the most ancient Arctic populations; in the Rhône valley, in search of strange cannibals; or on the slopes of Mount Ventoux, the giant mountain in Provence, to discover mysterious deer hunters – who in fact hunted exclusively stags in the prime of life, a hundred millennia ago, in the vast primary forest of Europe, before the coming of the last ice age. Along the way, I will examine the Neanderthal's view of the world and question ours. I will try to envisage its rituals of life and death. I will explore its ways of being in the world, which bring us back to our own humanity and the fragility of our own ideas. And the picture I will paint will be an original one: neither human nor ape, with its own way of being human that is not ours. These explorations, these thoughts, these discoveries, these questions, these researcher's hesitations are an invitation to a journey. A Homeric journey of body and soul, like every true journey. A journey to a distant place, of course – it is always a long journey – but you can remain seated, or rather on all fours, in among the rocks or on the banks of large rivers where in a frozen, fossilized state we find scenes, actions, events that speak to us of peoples far away, in both space and time. Peoples erased from our irredeemably amnesiac memories. Peoples for ever extinct.

Extinction

Because there was indeed an extinction. A final full stop.

Sudden, unexpected. And there it is, in front of us, like an enigma with no clues, but a head-spinning enigma. So even humans can die out without warning? The disappearance of a whole humanity, so near to us in time, should raise many questions. Can a humanity really become extinct?

Of all the questions I discuss here, this one is the easiest, so I answer it in these opening pages. Not only can a humanity come to an end, but this extinction is a clearly and definitively proven fact, even though geneticists have shown that there are still Neanderthal traces in the genome of the populations who are the current occupants of their ancestral territories. But these same studies have also shown that the Neanderthal was not genetically subsumed within us: the presence of a few genes revealing interactions with our ancestors cannot be interpreted to mean that this population somehow survives. These genetic traces are the marks of distant encounters between biologically diverse populations which were probably only partially fertile. Based on these traces some have suggested that this extinction was only a relative one – a type of dilution, in other words. Such suggestions are not only scientifically erroneous but also fundamentally specious.

Imagine for a moment that every species of wolf suddenly became extinct on earth. Farewell, *Canis lupus*. Now apply the theory of the genetic dilution of the Neanderthal in *Homo sapiens* to the case of the wolf. The result of this sulphurous alchemy would be to suggest that wolves are not really extinct since whole sections of their genes remain detectable in the genome of poodles, *Canis lupus familiaris* . . .

Well, the wolf has been luckier than the Neanderthal as it hasn't become extinct, but if it had it is immediately obvious that the poodle that survived it could not seriously claim to be the descendant of its remarkable cousin.

We are the Neanderthal's poodle. I don't mean by that that we are the prettied-up, domesticated version of the original wild beast, but that, just as the poodle is not a survivor of the wolf, so the Neanderthal does not live on in us. That humanity is extinct, totally extinct. That human branch is no more, and its essence, which we will explore together, is irreversibly extinguished along with it.

One might question whether this temptation to mitigate the greatest human extinction by assimilating it within a genetic dilution that did not take place has a whiff of revisionism about it. Might it not be a distraction from the striking correspondence between the expansion of *Homo sapiens* across Eurasia and the greatest recorded human extinction?

It is actually not too difficult to exonerate our ancestors, the colonizers of Europe, for the extinction of the Neanderthals, since the supposed relationship between these two events generally relies on the purely circumstantial evidence that they occurred within a common time frame. But time in prehistory is measured with a margin of error of plus or minus a thousand years. This error of one to two thousand years derives from the imprecision of the carbon-14 dating method. According to that methodology, you no doubt dined yesterday evening with Charlemagne on your left and Julius Caesar on your right . . . *Bon appétit* . . .

The archaeological evidence documenting this exact moment is in fact extremely scant and the dating methods are too imprecise to establish any relationship between the colonization of Europe and the extinction of its aboriginal Neanderthal populations. But if you have even a passing interest in the Neanderthals and their extinction, you will certainly have seen in the media a steadily growing flow of rather mind-boggling new theories on the processes that led to their extinction. Given this flood of information, you'd be forgiven for thinking that dynamic archaeological field research is radically updating our knowledge of the Neanderthal at a rapid pace. And if you are imagining great international scientific programmes undertaking large-scale research in caves to resolve this enigma, you might as well disabuse yourself now. There aren't any.

In France, still considered to be the number-one nation when it comes to prehistoric research, no archaeological operation has turned up a new Neanderthal body since the late 1970s, and there have been virtually no new complete archaeological sequences, with flints, bones and human remains, to update our knowledge of the final millennia of this population.

On the one hand, our tools have evolved remarkably with the explosion of biomolecular analyses; on the other, for more than forty years, neither programmed research nor preventive archaeology have allowed us to update the fundamental bases of our scientific documentation. The Neanderthal extinction is a simple fact: the disappearance of a humanity and its ancestral ways of living, replaced suddenly by the new era of the recent Palaeolithic, borne into Europe by the powerful waves of the *sapiens* influx.

Art Builds Bridges across the Ages

We need to understand the emergence of this new era, which announced itself like an icy breath through the silent death of the Neanderthal more than forty millennia ago. Does the recent Palaeolithic, the era of painted caves and ivory statuettes, seem distant to you like some confused dream of clashing rocks and grunts? You are wrong. Absolutely wrong. The era of this first *Homo sapiens* in Europe is the era we live in. Those humans are us, totally us. They are without any fundamental divergence every human society we have known since the beginning of their reign in Europe. Everything about these ancestors, from the fortieth millennium onwards, is familiar to us: their huge domestic architectures, real nomad cities of central Europe built in the skeletons of mammoths, their artefacts and the elegant, stylized lines of their statuettes of polished ivory. The symbols painted on the walls of their subterranean sanctuaries 34,000 years ago hold their own alongside the greatest masterpieces of the Renaissance or the Impressionists – Degas, Monet, Renoir and the rest.

All our societies devolve from this Palaeolithic art. Between it and us there is an organic, powerful, continuous link, transcending time, spanning dozens of millennia, as if time had no thickness, as if it were a mere anecdote with no real substance. Just a brief hiatus, no more, between the first drawings of our *sapiens* ancestors and the graffiti on our modern concrete underpasses. All artists from the nineteenth century to the modern day, the true artists, the transgressors, from Gauguin to Picasso, have picked up on this. They have put in words and forms and colours this shock that unites prehistoric art with primitive art, this shock that they can feel in their bodies, like raw evidence, via their highly developed artistic sensibilities.

Whereas we must admit that we have no idea what Neanderthal art might have been like, conversely, we know that *sapiens* art is a continuum.

From Lascaux to *Guernica* is just a single step. One step, yes, but the step of a stroller, not even a march, not even an advance. The Cubists, the Fauvists, the Impressionists merely rediscovered what had already been expressed dozens of millennia before them. They were all staggered to discover that across the world and across time *sapiens* art is a single, homogeneous art. André Derain, in his letters to another great Fauvist, Vlaminck, wrote in 1955: 'I am somewhat moved by my trip to London as well as to the National Museum and the exposition on Black art at the Musée de l'Homme in Paris. It is amazing, terrifyingly expressive.' And Picasso, emerging from the mother of all painted caves, Altamira in Spain, exclaimed in amazement or satisfaction, 'They invented everything!'

By what mysterious means are the arts able to defy the millennia and communicate with each other with such ease, building bridges between the ages, fully free, fully self-possessed from the very start? Dozens of millennia summed up, without explanation, in a unique sensibility, in a single look, in a single perceivable stroke of the soul?

These bridges over which temporality holds no sway radically encapsulate the first *sapiens* art, from prehistory to primitive art. The *Sapiens* are a continuum. It is only the veil of our education that superficially hides access to the keys to human universality, that of *Homo sapiens*, keys scratched from first the painted walls to our now totally anthropized and artificial world. All the keys to understanding all societies, since the dawn of *Sapiens*, are there, all around us, right before our eyes, which have been blinded by an excess of the obvious. Derain articulated the only real conclusion, perhaps, that can be drawn: 'What we need is to stay young, perpetually a child: we could make beautiful things all our lives. Otherwise, when we become civilized, we become like machines that are well adapted to life but nothing more!'

Goodbye, My Other Half

This evidence that meets our eyes despite the passage of time is the same that leads us to conclude that the Neanderthal could not have been *human* in the way we imagine.

With the Neanderthal, my dear old friend, there are no great rock frescoes that speak to us across the ages, no eccentric jewellery fashioned from ivory or deer antlers, no polished animal or human statuettes in coloured stones. There are beautiful stone tools, for sure, and beautiful artefacts, often sublimely rendered. But let's face it, how many human societies are remembered for their knives, their tools, their weapons?

None, right?

Maybe you have heard about Neanderthal cave art? Musical flutes carved out of bones by the creature? Beautiful bracelets of eagle claws or pierced shells? Superb headdresses made of feathers of birds of prey like those of the Aztecs or the Lakota? If this has piqued your curiosity, don't hang around, jump straight ahead to the chapter dealing with Neanderthal art, but be prepared to abandon your preconceptions: while the creature displayed a remarkable sensibility, it was nothing like ours, and in the pages that follow we will trace the subtlety of such exotic sensibilities, which remain to be explored to this day.

Neanderthals are probably not versions of ourselves. The common features of all humans in the time since humans have been human don't seem to apply to our creature. Not only is it different, it is also extinct. And it did not merge into us, diluted in our genes like a sugar cube dissolving in hot water. Its genes in us are so rare and so unevenly distributed in the human population that we can now be certain that the key to this extinction does not lie in an improbable tale of cannibal love where the disappearing human provides the matrix for the new humanity. This idea that some of us are the inheritors of a disappeared humanity is a bizarre illusion. In reality, cross-breeding and hybridization between different species are commonplace in the natural world. There are felids, canids, ursids, suidae, and they are usually interfertile; we are surrounded by tigrons, ligers, boar-pigs or pig-boars, but these biological chimerae do not in any way illuminate the destiny of lions, tigers, pigs or wild boars.

What a strange idea, what a paradox, to link a human extinction to a love story, a total, absolute, cannibalistic form of love in which one is dissolved in the other. True, it's a nice story, easy on the ear. No human extinction, but rather a loving fusion: $1 + 1 = 1$. Goodbye, my other half, I loved you . . .

And to think that we poor researchers, we poor archaeologists, do not even know if the two humanities, the living and the dead, ever actually crossed paths on the immense European territories of Neanderthal aborigines, on the very site of their extinction. We're back at that dinner table alongside Caesar and Charlemagne, and that's as much as we know.

There are virtually no archaeological sites in the whole of Europe for which our measurements of time are sufficiently precise to establish with any certainty that the two humanities encountered each other. The victim has been identified, but we don't have the body, and we don't know the identity of the murderer – we don't even know whether the victim ever encountered his purported murderer.

At this stage, ladies and gentlemen of the jury, bring down the curtain, there's no case to answer, release the accused. On the whole, the specialists are more than happy to release the accused. Some of them go as far as to claim that our *sapiens* ancestors might have settled on lands already completely abandoned, devoid of any human presence for centuries if not millennia . . . So it's impossible to determine whether any crime, or genocide, was committed because this is simply invisible from an archaeological perspective. The perfect crime. Of course there is an obvious motive – colonization, which is beyond dispute – but where is the smoking gun? Even the bodies of the victims have never been discovered. And the alibi is watertight: there is no clinching proof of the two humanities ever encountering each other in any part of the European continent.

Let's be clear, this triple exemption isn't any such thing and tells us less about the real facts than about the dismal quality of archaeological recordings in the face of events that took place more than forty-two millennia ago. So could the colonization of the European continent by *Homo sapiens* provide both the motive and the process of this human extinction?

The question can't be ducked, given what we know so far. It seems, in fact, that if we look very closely using the most recent methods, the archaeological, genetic and also chronological evidence allows us to demonstrate that an encounter did take place. And we might even go so far as to raise the veil on the types of relations that some of these human groups were able to form. The motive can be rationally established, and the alibi crumbles.

We still don't know what actually happened. To get closer to the truth, we have to rely on the most solid and precise archaeological research, conducted on targeted archaeological sites over a long time period to a high degree of resolution. It is also necessary, so as not to introduce bias, to take enough of a step back from our own intellectual constructions and the authorized knowledge of the Neanderthals in academic circles.

To reach the naked Neanderthal, we must be honest enough to strip away all the tat with which we have decked it out. We have to return to the sources, focus on the structure of Neanderthal society, their craft work, their choices, their way of inhabiting the world, in order to see them purely in relation to the archaeological facts. And even as we set the train back on the rails, we realize that the weight of all the doubts, the questions, the uncertainties is already so heavy that it feels frustratingly as if the creature is refusing to be analysed, to be too easily categorized.

Do we even know all the environments that these populations succeeded in colonizing? When we look at the fringes of the world, in the polar regions, we realize that even this simple question is astonishingly difficult to answer.

2.

A Boreal Odyssey

A World of Ice

Now you know that no one should talk about the Neanderthal without getting right up close to it, in the places where it lived, in the wild open wastes and the recesses of cliffs where the subtle remains of its material existence were fossilized – and certainly not in the drawers of a museum. Nevertheless, it is in a drawer that I will begin one of the first stories connecting me to these extinct societies that I wish to recount to you. A drawer in the Institute of the Ural Branch of the Russian Academy of Science, in the boreal city of Syktyvkar in the Komi Republic, the small Arctic republic at the north-east corner of Europe. All the archaeological sites that inform us about the first populations of the Arctic region have been discovered in the vast open spaces of Russia. This, strange though it may seem, is where we begin our research into the Neanderthal: high on the European Arctic Circle.

It is this Arctic climate that is most characteristic of the environments in which Neanderthal societies developed on the European continent; we have to go back more than a hundred millennia before the climate archives record much more favourable temperate conditions and temperate global climates which were even, for a few dozen millennia, much warmer than modern terrestrial temperatures. Before the ice and the vast grassy steppes, temperate Eurasia was covered by a vast primary forest, effectively limitless, infinite expanses in which no tree was ever cut down. This was on a scale that is beyond our imagining, and in the following chapters, to warm ourselves up, we will confront these peoples of the forest, these Neanderthals of the woods, whom scientific research is only just beginning to properly identify.

At this time, Eurasia was covered by icy wastes, and the Neanderthal, for the dozens of millennia up to its extinction, was a polar creature. But there is polar and there is polar – and archaeological research shows that a very small number of Palaeolithic societies colonized the boreal spaces of the Russian Arctic even when the planet was plunged into the deepest ice age.

During this climatic phase, the earth's temperatures crashed. For dozens of millennia the three Scandinavian siblings Norway, Sweden and Finland were frozen, covered by massive ice sheets. In the very coldest periods, the fronts of these immense glaciers advanced and covered most of Britain and Ireland, leaving only the extreme south of the British Isles free from ice. So the sea level was much lower, with huge quantities of water locked up in the ice. The English Channel was not a maritime strait but a wide valley with a river flowing through it into the Atlantic well out to the west, somewhere between present-day Brittany and Cornwall.

It is not out of the question that Palaeolithic populations ventured into the icy wastes of northern Europe, but so far we have found no archaeological evidence for this. Surprisingly, a bit further east, in the present-day Komi Republic, the northern regions were never covered with ice, even well above the Arctic Circle. The broad Pechora river, which flows out north into the glacial Arctic Ocean, was for a while blocked by masses of ice and formed a gigantic lake. But the colossal pressure of the water released the jam, definitively liberating the polar regions, which then remained ice-free, as indeed did the boreal expanses of Siberia. How can we explain the fact that in these ultra-continental areas, which today are among the coldest polar regions in the northern hemisphere, no ice sheets ever developed during the last ice age? The reason for this paradoxical situation appears to be fairly simple. The powerful glaciers that covered Europe from Ireland to Finland created a natural barrier, cutting off the continental polar regions from the Atlantic Ocean. The precipitation, which came essentially from the Atlantic, was harvested by these vast expanses of ice and never got further than this glacial block.

The polar climate of the north of Eurasia was very cold but very dry and so stayed free of the grip of ice. Not only did the land remain uncovered, but in the milder season it offered a biotope that was exceptionally and particularly propitious for life. In the polar and Siberian expanses herds of proboscideans developed in large numbers, creating an environment so unique that today we call it the mammoth steppes.

Living in the Cold, Living from the Cold

There is a piece of Inuit wisdom that might be summed up as follows: cold is never a problem for humans. Access to proteins and staple foodstuffs is the only factor that limits human expansion. Our bodies are not very susceptible to cold when it is dry, and these Eurasian polar regions of the last age were markedly dry, probably even arid, in their coldness. In 'feels like' terms today, it is colder in St Petersburg in February at -16 °C than in the continental expanses of Siberia at -30 °C. During my research in the polar regions, I had to experiment with the reaction of my own metabolism when, for weeks on end, I was daily exposed to temperatures of -25 °C. After a few days, probably less than ten, I found that my body no longer suffered from the cold and I could spend the whole day walking in the taiga without ever really enduring the experience of cold.

My metabolism had swiftly recalibrated and adjusted itself in a matter of days, to the extent that the temperatures just seemed normal to me, even quite pleasant. Even more surprisingly, I worked on different projects from the Sahel to the Gobi desert to the Horn of Africa where I usually experienced torrid temperatures. My metabolism worked in such a way that, even in extreme temperatures, my body did not sweat, or at least very little. And yet in the middle of February, in the European polar regions after walking all day in the snow, when I got back to my apartment, which was heated to between 18 and 20 °C, I started sweating, sweating to the tips of my fingers. My metabolism was calibrated to -25 °C, a temperature that didn't bother me in the slightest, but I found the more familiar temperate atmosphere stifling. It shows how well our bodies can adapt, which has important implications for our perceptions of human expansion in boreal regions.

It is common to read, even in the work of very good researchers, that the colonization of even the middle latitudes of Eurasia by populations emanating from African environments probably required technological adaptation, developing techniques of protection against the cold, as well as social adaptation and establishing robust mutual-aid networks. They say it was thanks to these technological developments and the singular organization of their societies that our ancestors were so successful in conquering the most testing environments and climates on the planet. These theories attribute a central role to the inventiveness of humans and their strategies in compensating for their metabolisms, which were more specifically adapted to tropical climes. This is probably a preconceived idea that takes no account of the amazing biological properties of the human metabolism. It is likely that such concepts are erroneous and tell us nothing about our biological reality or about the precise organization of those distant Palaeolithic societies. These perceptions, these views, even if scientific in their approach, may well be, above all, prisoners of conceptions of the world and of human beings that do not apply to distant prehistoric societies but which say more about ourselves, modern-day Westerners, and our inability to project ourselves into realities that are quite strange to us.

It was on this very basis that, around the turn of this millennium, various theories about the extinction of Neanderthal populations emerged. Based on the absence of Neanderthal sites north of the 55th parallel, researchers put forth the hypothesis that these populations would not have been able to adapt to high European latitudes, limited as they were by their technologies or their inability to construct mutual-aid networks that would enable them to surmount the most extreme environmental constraints. The Neanderthal populations would only have been able to colonize the middle latitudes and would not have been able to deal with the climate changes that affected their biotopes in the last few millennia of their existence. The extinction of the Neanderthals, then, was the consequence of a simple change of climate and an inability to adapt to new biotopes.

These different hypotheses of the extinction of the Neanderthals focus on environmental factors and see the impressive expansion of *Homo sapiens* across Eurasia as only of secondary importance. Whether considered individually or collectively, these propositions seem very shaky. Can anyone really believe that the Neanderthals merely evaporated like snow in the sun?

The evidence of colonization at very high latitudes calls into question climate-based theories and the idea that they had limited adaptability. Human metabolisms do not react like those of plants when faced with changes of climate. Human bodies show themselves to be remarkably adaptable, ubiquitous and capable of easily dealing with the whole gamut of earthly environments. As we explore the confrontation of archaic human populations with polar environments, we rely on our specific conception of our own ability to adapt. This is what Wim Hof – 'the Iceman', as he is known – has taught us. In the winter of 2007, Wim Hof ran a half-marathon of 21 kilometres on the Arctic Circle, wearing nothing but shorts. A few months later, he tackled Everest from the Tibetan side without any equipment to protect him from the cold. Wim Hof is now a genuine case study for researchers seeking to understand the human metabolism. One of the lessons of Wim Hof, who is no superhero, just a man of flesh and blood, is that the human body can adapt very well to the cold, that our metabolic properties are probably not determined by the question of our biological origins, which we know are African and tropical.

So there we are, very probably trapped in our own projections, fantasies, fears – natural fears, of course, but ones which are resistant to experiment. Palaeolithic societies probably did not need special technological or social means to deal with the diverse biotopes of the planet. Their bodies alone probably did most of the work.

Great Polar Expanses

The polar regions offer a remarkable key to understanding the organization and structure of distant Palaeolithic societies. In

2006, I went to western Siberia to present my research on the final Neanderthal societies at the Nordic Archaeological Congress. This adventure would eventually lead me to the western and Siberian flanks of the polar Urals, following the traces of the very first boreal populations. As you cross these immense wastes today you come across dachas, polar taiga and former gulags where, more than anywhere else perhaps, a certain melancholy of the Slavic soul has taken up residence, embedded in the concrete blocks, washed up in the vast industrial ruins, the rusty corpses of Soviet idealism. These carcases of iron and stone had few redeeming features for me, but there is a profound humanity here, both touching and disturbing. I wanted to experience it too. And boreal spaces had always attracted me. Was the Neanderthal a polar creature? Hadn't it spent most of its existence in the throes of the last ice age? What did the ancient societies of the Palaeolithic come here to do in the polar regions during the most challenging climatic phases recorded on earth in a million years?

The top Russian specialists of the Nordic societies had gathered for a few days in Khanty-Mansiik, in the Siberian west, for this Nordic conference. It was the end of September, and the first snows were starting to cover the banks of the Ob, one of the huge rivers of the north, its immense size on a scale commensurate with that of Siberia itself. A landscape unlike anything to be found in western Europe. The Ob crosses the whole of Siberia, from north to south; its delta alone, covering three million square kilometres, is almost as large as that of the Nile, the longest river in the world, the equivalent of almost five times the surface area of France. These outsize dimensions are just a numerical expression of the vast, wild open expanses that originate on the European flanks of the Urals and only come to an end on the distant shores of the American continent.

In the middle of the twentieth century, Russian researchers established a pioneering school of Palaeolithic archaeology, setting up research strategies that were imported only much later into western Europe, by André Leroi-Gourhan in particular, a man of rare intellect whose interests encompassed archaeology, ethnology and philosophy. Leroi-Gourhan had been profoundly influenced by the great programmes of archaeological research developed by the Soviets. Russia is largely covered with loess, thick layers of silt deposited by the winds, which fossilized at very high speed the habitats of Palaeolithic hunters, preserving the remains of vast camps of nomadic populations. In these very extensive archaeological reservoirs the Soviets developed ambitious methods of wide trenching at sites abandoned by Palaeolithic hunters to reveal soil strewn with flint tools sometimes embedded in veritable mass graves of mammoth bones. Soviet research programmes have made an indelible mark on global archaeology, even if they are happily no longer driven by the megalomania of the Soviet system.

Nowadays, Russian research, the inheritor of this exceptional archaeological programme, is still remarkably dynamic, but for the archaeologists the task here is virtually insurmountable. How do you manage and preserve the entirety of a heritage which covers an area from Europe to America? Russia is by far the world's largest country; its population, which is only a little more than twice that of France, occupies an eighth of its total surface area. Russia is roughly twice as large as the next-largest nations, Canada, the USA and China. Consider that half of this land is covered in immense primary forests. These boreal expanses represent more than a quarter of the world's forests and constitute the largest arboreal wilderness on the planet, far surpassing the equatorial forests. The human population is largely concentrated in a few large cities, like human colonies in oceans of virgin greenery in which forestry has made but small inroads compared to the total surface area. Siberia, especially its polar regions, represents, along with the Antarctic, the last place on earth that has been wild for ever. There is still an untouched Wild East just as there once was an immense Wild West. This is without any doubt the final frontier.

A Race against Time

Given this massive scale, how do you manage a colossal archaeological heritage buried under infinite expanses of loess? In the higher boreal regions, vestiges have been preserved in frozen soil for thousands of years. But in these latitudes accelerating climate change is causing the soil to thaw in front of our very eyes, liberating their archaeological treasures. Flesh, wood, leather, woven materials, basketry, nets are re-entering the cycle of putrefaction from which they have been spared since the ice age. Mammoths and rhinos have been discovered by the sparse inhabitants of the Siberian expanses, trappers or reindeer herders of the far north, but what has happened to the bodies of the Palaeolithic hunters in this wilderness? Incredibly, the greatest finds were made by a child playing on a riverbank and an artist searching for ivory for his marvellous sculptures. Just like that, by the side of a stream, in a marsh, as the remains emerged from just-thawed prehistoric ice. There is no reason why the suspension of the effects of time by freezing should affect only wild animals and not human remains. It is likely that these Palaeolithic bodies have already emerged from the ice and returned to their cycles of decomposition. We must also consider the possibility that a certain number of bodies from the distant Palaeolithic have already been found by the local people and that perhaps their fate was to be given a decent burial where they were found or in the nearest cemetery. If this is the case, they lie today beneath a cross made of spruce or larch.

With the thawing of the boreal soils, frozen in ice since the last ice age, we are confronted with a cruel paradox. The melting of the permafrost reveals concealed archaeological sites while also bringing about their rapid and inexorable destruction. In the vast Nordic lands the sites are invisible, buried under layers of loess sometimes dozens of metres thick. As there are no roads, it is impossible to bring in mechanical diggers and bulldozers. The sites are generally only revealed by the natural erosion caused by the larger Siberian watercourses. The current releases bones and flints from banks that are carried away by the waves below at the edge of the riverbed, revealing the existence of sites that have been sealed for tens of thousands of years. Once the sites are revealed, they have only a few seasons before they are swept away by the powerful currents of these vast rivers, often in spectacular fashion.

My Russian colleagues have experienced this strange phenomenon. Their teams took advantage of the few weeks of good weather in the Siberian polar north to dig out some remarkable archaeological fossils held in the ice for nearly thirty millennia. The site had been revealed by erosion by the river Yana in eastern Siberia and stood several metres above the banks of the river. When they got back from their lunch break, the whole site had simply disappeared. A 20-metre cube of frozen soil had just collapsed in one piece into the river. The archaeologists then had a real race against the clock and had to carry out their work in extreme circumstances. Polar sites are miles from the nearest human habitation. Nothing here is easy. You can only get there in the first place by boat or helicopter. You have to camp, deal with the local wildlife – wolves and polar bears – and then, during the short temperate summer period, try to uncover some archaeological remains. But these vestiges are held in frozen soil and cannot simply be extracted using metal tools: you have to melt the ice. The classical procedures alternate between high-pressure water jets and the teapot approach, which consists of delicately pouring warm water on to frozen ground to release the precious archaeological vestiges.

Such missions are physically demanding in these high latitudes, despite the ever milder climate, which allows easier extraction of certain archaeological sites that have previously been inaccessible. The thawing soil allows us to locate and access previously unapproachable archaeological sites and at the same time sets in train their irreversible destruction. Not over the course of a century. Not even a decade. But in front of our eyes, in real time. We cannot monitor the whole of the wild Siberian expanses. Season after season, the destruction of this heritage continues on an almost daily basis, and we have to recognize that we are powerless to do anything about it. Our access to the Palaeolithic populations of the Arctic regions is only a very fleeting one.

On the Trail of the First Polar People

Imagine this: despite the impressive efforts of Soviet archaeologists, there are only three recognized Arctic archaeological ensembles older than twenty millennia on earth. These three sites are all in present-day Russia. Two of the sites were discovered in boreal Europe, not far from the western flanks of the Urals. The oldest, Mamontovaya Kurya, is 40,000 years old and is situated right on the Arctic Circle. Only seven carved stone tools have been found there, alongside an enormous mammoth tusk regularly incised along its full length. The incisions are very deep and were made using a stone tool. The object is still a mystery, and these marks have no exact equivalents in the Palaeolithic records of Eurasia. Are they decorative, a method of counting, or simple chopping marks? It is difficult to answer these questions or to attribute any real decorative value to them. Mamontovaya Kurya is accessible to archaeologists only during a few weeks of the year, and in some years only for a few days, when the level of the river Usa is low enough to allow access.

The site is buried under 18 metres of sediment, but erosion by the river has opened up localized access to prehistoric vestiges requiring the removal of only 4 to 5 metres of sand, throwing up bones and objects discarded 40,000 years ago by Palaeolithic settlers of the boreal regions. Some large-scale archaeological excavation has dislodged around 50 square metres of ancient soil, but only this handful of seven tools, made of flint, schist and quartzite, has been discovered. The carved objects were usually found alongside mammoth remains, but there were also bones of reindeer, wolves and horses. It seems likely that these four species were animals hunted by the first polar populations, but there are too few remains to be confident that this collection of bones is from the hunts of the people who lived here by the river Usa.

There is almost nothing more to say about these enigmatic early settlers of polar Europe. Analysis of the very rare objects they fashioned does not allow us to determine with any certainty whether they were inheritors of the ancient technical traditions of the Neanderthals or whether the vestiges were abandoned by modern humans, colonizing the polar regions more or less at the same time as the rest of the European continent. It was precisely around this fortieth millennium that the two populations could have coexisted in Europe. So we need to look at the other two sites to try to better understand the dynamic of the colonization of the boreal wilderness. But before we turn our attention to these other two ancient sites of the polar Palaeolithic, we should consider some remarkable evidence of human presence in boreal lands some distance away. Evidence that would push colonization back eight millennia before Mamontovaya Kurya. Did humans in fact cross the Arctic Circle long before the fortieth millennium?

The Invisible Polar Hunters of the Forty-eighth Millennium

Though there are only three polar sites containing objects fashioned by the human hand, in 2016 the American journal *Science* announced the discovery of a mammoth carcass bearing traces of butchery using stone tools. The mammoth was discovered on the Taymyr peninsula, some 600 kilometres north of the Arctic Circle. The Taymyr peninsula is located on the top end of the Ural mountains, in the far north-west of Siberia. It is the most northerly region in Eurasia and it covers an area larger than the whole of Scandinavia. The peninsula has only a few tens of thousands of inhabitants, largely grouped around a handful of mining towns that sprang up in Soviet times, squares of concrete floating in the middle of the boreal taiga. The inhabitants are spread very thinly across this vast territory: Dolgans, Nganasans, Nenets, nomadic people whose livelihoods depend on hunting and herding reindeer and who live in yurts, the sumptuous tepees of the Siberian peoples. This is essentially virgin territory, in archaeological terms, and the mammoth that interests us was discovered by chance by Evgeny Solinder, a schoolboy who was playing on the banks of the river Yenisey.

The Yenisey rises near the Mongolian border. It crosses the whole of Siberia before flowing another 5,000 kilometres north into the Arctic Ocean. It is another Siberian super-river, like the Ob. The mammoth vestiges lay a few hundred metres from the polar weather station at Sopochnaya Karga. A swiftly assembled archaeological operation managed to extract the whole of the pachyderm's remains in a single block, which was frozen and sent by cargo plane to the research institute in St Petersburg. The mammoth was remarkably well preserved and still retained traces of flesh and fur. Analysis of the remains showed marks clearly made by humans. The marks were indisputably inflicted by stone tools and showed that the animal was slaughtered and its flesh removed, including its tongue, behaviour well documented among other Palaeolithic hunter populations in more southerly regions of Eurasia. But no tools or traces of the hunters were found in the vicinity of the carcass. Carbon-14 dating showed that this mammoth lived no later than 48,000 years ago. Such an astonishing discovery proves that humans occupied the boreal regions far beyond the Arctic Circle at a time when *Homo sapiens* hadn't even begun to colonize the European continent. No other archaeological site that was this old has ever been found in the Arctic region, whether in Siberia or in Europe. It is impossible to comprehend the significance of these hunting vestiges. The archaeological traces of these distant polar populations are still completely invisible to us to this day. Not a single carved stone tool, not a single piece of material evidence of these first boreal settlements; they remain a complete archaeological enigma. The carcass bears witness to a polar mammoth hunt by a people who are totally unknown to the scientific community.

Two thousand kilometres further east, the shoulder bone of a wolf was discovered in turn above the Arctic Circle in Yakutia on the banks of a tributary of the river Yana. Analysis of the bone showed wounds inflicted by manufactured weapons – spikes, spears or arrows. But the wolf did not die of its injuries and managed to escape. Dating of a bone in the wolf's front left paw gave results identical to that of the mammoth hunted on the Taymyr peninsula and pointed to a hunt which took place forty-eight millennia ago. We are once again well to the north, but also . . . to the east. Our poor Siberian wolf was no more than 2,000 kilometres from the Americas, the exact same distance as from the mammoth of Taymyr. The most boreal regions of the planet were thus colonized in very ancient times, potentially much earlier than the major phases of colonization of the higher latitudes of Eurasia by *Homo sapiens*. We see here the trace of the hunters, but no tool offering us the slightest direct evidence of these peoples.

However, the wolf was found just a few kilometres from the river Yana, at the spot where one of the three documented polar sites anterior to the twentieth millennium was discovered. This is the Yana RHS deposit.

An Unknown Boreal Civilization Frozen in Ice

RHS stands for 'Rhino Horn Site', one of the earliest discoveries that introduced us to these unexpected archaeological ensembles. Here, 500 kilometres north of the Arctic Circle, the archaeologist Vladimir Pitulko uncovered exceptional remains of Palaeolithic encampments, frozen in ice for thirty millennia. Unfortunately, the Yana remains date from a period fifteen to twenty millennia after the subtle evidence of boreal colonization provided by our mammoth and our wolf.

Vladimir Pitulko, an expert in the populations of the Siberian far north, has conducted digs on a large scale across the Siberian east and uncovered polar sites where the technologies show little resemblance to those known elsewhere in the more southerly latitudes of Eurasia. An amazing discovery, then, virtually unforeseen, and rich in archaeological data. We are in the delta of the river Yana, a very modest Siberian river, having a basin only twice as large as the biggest river in France. Most importantly, the river is only 872 kilometres in length and so provides no link with lands at lower latitudes, as its source already lies within the sub-Arctic zone.

The Yana sites have yielded tens of thousands of carved stone tools, but also very refined *objets d'art* in mammoth ivory displaying great expertise in the handling of the materials. Research showed up 1,500 pearls carved out of ivory, but also fox canines and reindeer incisors. Reindeer antlers were chopped up into pieces and then carved into small animal figurines. Small, square-shaped and finely decorated ivory bowls were uncovered, but also bracelets and other pieces sculpted out of ivory and interpreted as tiaras and diadems. There was a profusion of these delicately crafted objects, some of them decorated by fine geometric markings. Prior to this, no manufactured objects had been discovered apart from the seven stone tools from the fortieth millennium found at Mamontovaya Kurya; Yana RHS would yield up thousands of individual pieces displaying a complete mastery of crafting objects from all sorts of raw materials, from ivory to flint, via reindeer antlers and rhinoceros horns.

A fully mature polar civilization, fully adapted to this singular environment and unknown in the archaeological registers, had suddenly emerged from the boreal wastes. Thirty thousand years ago, these hunters were able to take advantage of a wealth of fauna, including mammoths, woolly rhinoceros, bison, reindeer, brown bears, wolves, gluttons, muskoxen, horses, polar foxes, hares and ptarmigans. By all evidence, in the middle of the ice age, these people of the far north not only survived but thrived in environments that to us seem extreme, but which in fact contained remarkable riches. The excavations revealed remains of thousands of hares, which inform us that Leporidae were trapped in a systematic fashion. Piles of their carcasses were discovered, left where they lay, uneaten. The people of Yana showed no interest in their flesh but exploited small game animals only for their skins: soft, warm, but also very fragile. Jean Malaurie, the French polar explorer, who lived in Greenland with the Inuit of Thule in the 1950s, recounted that the small community with whom he spent a few seasons might hunt up to 1,500 hares each year just for their skins, as they considered their meat rather insipid. Great polar minds think alike across the millennia.

The tundra steppes were treeless, so these people had to adapt their technologies to the absence of wood, an essential material used across Eurasia for hundreds of thousands of years to make hunting weapons, spears, javelins and arrows. Shafts such as these were absolutely necessary for hunting the reindeer, horses and bison that inhabited the polar spaces. At Yana, they ingeniously used ivory instead of wood. The hunters would kill mammoths primarily for their tusks. Apart from the tongue or the meat of young mammoths, which was probably more tender and tasty, they seemed to have little interest in the flesh of the pachyderms. So the people of Yana concentrated on hunting females. Why females? Because their tusks were much straighter than those of the males. The hunters would expressly seek out these large, less curved tusks in order to sculpt them into the javelins they needed for hunting, which they had no other way of procuring in boreal wastes.

Who were these people of the far north? Based on the archaeological collections, their identity is in little doubt: their technologies are clearly modern technologies which to this day are considered exclusive to our biological species. They are certainly original and quite dissimilar to what we know of more southerly Palaeolithic populations in Siberia or in Europe as a whole, but there is no real confusion here: these technologies, the knowledge they represent, are undoubtedly those of a *sapiens* people. Everything here points to a modern society: the thousands of pieces of jewellery, the figurines, the decoration, the ivory javelins and even the fine ivory sewing needles, a subtle but fundamental technology for making finely woven clothing offering perfect protection against the cold. While they display the level of knowledge of a modern society, such technologies are profoundly original in some of their decorations but above all in the precise technical skills wielded by the artisans, and have relatively little in common with the work of more southerly populations.

Large areas of the Yana site are opened up each year by archaeological excavations in order to try to elucidate these enigmatic polar colonizations. The digs finally turned up two human milk teeth, which were immediately subjected to genetic analysis. These teeth, which had been preserved for thirty millennia in the frozen ground, have retained much of their genetic information. Not surprisingly, the DNA proved that they belonged to a group of modern humans but nonetheless revealed a population previously unknown to genetic science. This boreal civilization was distinct from the other Palaeolithic populations already recorded by genetic means, a now extinct population that the geneticists dubbed the 'Ancient Siberians of the North'.

They also bore certain Neanderthal genes, as do all the current populations of Eurasia, but their sequences of Neanderthal DNA were much longer than those we see in modern-day people, indicating that the encounter between the two populations was much more recent. It might have occurred between eighty and a hundred generations before the birth of the children who lost their milk teeth at Yana. And, rather remarkably, the Ancient Siberians of the North showed no sign of interbreeding with the Denisovans, another fossil humanity cousin of the Neanderthals known to have lived in the Altai, the mountains in southern Siberia. Yet the genetic imprint of the Denisovans is widely distributed among the present-day populations of south-east Asia and even as far as Australia. The origin of the Ancient Siberians of the North is a mystery to us, as is their eventual fate, but before they colonized these polar regions, or even during the course of this colonization, the ancestors of these enigmatic populations crossed over with the Neanderthals. They bore within them the genetic memory of these extinct populations.

The signature of an encounter with our Neanderthals, combined with the absence of Denisovan genes, might suggest that such an encounter took place a long way away from the middle and lower latitudes of Asia, where the genes of the Denisovan populations are still commonly found. Therefore further west, towards the European continent, and potentially further north, towards the polar regions that these populations would rapidly colonize. It is even possible that the genes inherited from the Neanderthals, a population believed to be biologically well adapted to cold environments, might have provided an advantage for the Ancient Siberians of the North in their successful colonization of the boreal regions. But more than simple biological attributes providing better adaptation to the cold, surviving in the harsh Arctic environment, as Inuit history tells us, requires a familiarity with that environment and carefully planning resources according to the needs of each season. Only an understanding of the particular conditions of these high latitudes would get you through the long and challenging winters of the polar climes. Unfortunately, the fauna extracted from the permafrost at Yana had no flesh or fur still on it. How is that possible, given that the remains were abandoned here in the middle of the ice age and their vestiges should still be difficult to extract from frozen ground even in our day?

A Polar Eden

Global climate archives show us that past climates were characterized by profound fluctuations of temperature which set the rhythm of our biosphere, a regular alternation between temperate phases and glacial phases. The origins of these climate changes are still little understood, and so they are not always easy to model precisely. We can't tell when the next ice age will arrive. Still, it appears impossible that we will be able to escape the next glaciation, as natural cycles include rapid and very powerful climate change. Such major fluctuations seem to be connected to periodic modifications in the earth's orbit. For dozens of millennia our planet has been subjected to polar conditions which have a bigger impact at continental high latitudes than in regions at middle and low latitudes, where 99 per cent of the current world population lives.

We are currently in one of the temperate phases, which began around 11,700 years ago, bringing the last ice age to an end. This deglaciation went through a number of phases, some of them recording remarkably abrupt climate changes, especially in the boreal regions. In Norway, we now know, thanks to the work of the Norwegian glaciologist Jan Mangerud, that at the start of this warming the glaciers in the Scandinavian fjords retreated more than 160 metres a year. In ten years, these vast millennial glaciers would retreat more than 1,500 metres. As the deglaciation was perceptible within the span of a human lifetime, it was a very concrete experience for human societies, a common fact of life for boreal societies, who physically witnessed the transformation of their territory in front of their eyes, year after year. This climatic upheaval was global, impacting all the world's ecosystems and causing sea levels to rise by more than 60 metres.

We are currently living through an interglacial phase, the Holocene, which, at our latitudes, gives us incomparably more favourable conditions for life and biodiversity than at any time in the preceding tens of thousands of years. It was at the start of the Holocene that the global climate was the most temperate. The summer temperatures in the Arctic zones were in fact markedly higher than those that we know today and set off large thaws of the permafrost in the polar regions where the ground is currently frozen. The Yana RHS deposit, trapped in the permafrost, has not delivered any of the perishable organic matter – skin, flesh, hair – preserved elsewhere in frozen Siberian soils, which regularly throw up entire mammoths or rhinoceroses, with meat, fur and intestines still full of their last meal. This organic matter was lost at the start of the

Holocene, when some of the ground underwent a complete thaw, leading to the decomposition of the most fragile archaeological remains. At 500 kilometres north of the Arctic Circle, the current zones of permafrost underwent a thaw which lasted several centuries before freezing again in the progressive drop in temperatures which has been affecting world climates for more than six millennia.

This trend is now in reverse due to human activity, which is interfering with the natural cycles of the biosphere before our eyes. The countless climatic changes that will play out on earth in the millennia to come will interfere with human activities, but they are unfortunately not predictable by any modelling. On much shorter timescales, we see the deglaciation of polar areas playing out rapidly in front of us as they gradually give up the carcasses of the great fauna of the Palaeolithic.

We are all familiar with the striking images of perfectly preserved whole mammoths which emerge every year from the soil of the Siberian far north, but this wilderness also regularly throws up remains of numerous other species: woolly rhinoceros, muskox, bison, brown bear, horse, reindeer, polar fox, wolf and more. So the polar biotopes had an infinitely greater biodiversity than can currently be recorded at these latitudes. And the density of the animal populations appears to have been remarkably high. Could it be the case that in these zones of the far north the temperate periods were more inhospitable than the glacial periods?

Analysis of the mammoth steppes reveals an extraordinarily rich environment, which gives us a striking, almost paradoxical, image of the Siberian biotopes of the last ice age. To accommodate the large Arctic fauna, the polar landscapes must have been covered by enough lush vegetation to sustain the needs of these large herbivores. A modern-day elephant eats between 60 and 300 kilos of grass a day, which gives us a scale of needs for the geographic polar expanses commonly called mammoth steppes. It was a rich milieu, then. Very rich, even: grasses and sedge, and a lot more. We know from analysis of pollen fossils and thanks to a study of the DNA of the soil that in springtime the Arctic and sub-Arctic regions could be carpeted in flowers: anemones, poppies, buttercups. The polar regions, with their herds of very large herbivores, were very different from the marshy tundra that we can admire today.

Were these regions of the far north particularly hospitable to life at the peak of the ice age? It is a counter-intuitive idea. Even at the lower latitudes of temperate Europe, this ice age was characterized by the extreme severity of its climate. Could it be, however, that development of human societies beyond the Arctic Circle reveals the existence of environments replete with game and particularly conducive to the development of such remarkable boreal societies?

It is possible that associating the colonization of the polar regions with social or technological prowess is a misinterpretation. A social construction. Our social construction. The construction of a point of view, as subjective and relevant as a 'It's warm' or 'It's cold' today. Pictures of Inuit children playing naked inside igloos might make us smile. When it is -45 °C outside, then 0 °C inside the igloo is warm. How is this possible? The human body is not a thermometer but an amazing machine that can recalibrate itself to the surrounding environment in a matter of days. This is not just a poetic image, glorifying the amazing adaptive qualities of the great apes that we are. It is not mere words: we can physically experience this phenomenon of our body regulating itself without our being aware of it.

A Refuge on the Edge of the World

This memory takes me back to February 2007, when I was working through the drawers of the Academy of Sciences at Syktyvkar, in the Komi Republic. Earlier I hinted at this research project, which started with the opening of a drawer. Well, here we are. The drawer contained collections from the digs at Byzovaya, the last of our three ancient polar sites on earth.

At Byzovaya we are back on the western flanks of the polar Urals. Back in Europe, then, in one of the most remote regions on the continent. My Siberian conference in 2006 had led to an invitation to work on the Palaeolithic collections of the European far north. These archaeological collections were created thanks to the research of Pavel Pavlov of the Ural division of the Russian Academy of Sciences. Pavlov had been conducting archaeological missions on the western slopes of the Urals – from the sub-Arctic territories to the Arctic Circle – for years, which provided the basis for collating all the early data for the colonization of boreal regions of the continent. The data were absolutely original and did not seem consistent with our existing knowledge of the Palaeolithic peoples of the same era gathered at more southerly latitudes. The collections assembled from these years of research were deposited in the Komi Republic. In Eurasia, four countries have polar territories: Norway, Sweden, Finland and Russia. The latter is divided into twenty-two republics, three of which are in the polar regions, their names evocative of the adventure of travel: Sakha, Karelia, Komi.

You may have never heard of the Komi Republic . . . A small area in Russian terms, it is slightly larger than Germany, and bisected by the Arctic Circle. Covered mainly by vast tracts of primary forest, it contains the largest forest still in a wild state in the whole of Europe and is classified as a World Heritage Site by UNESCO. Its capital, Syktyvkar, is situated 1,000 kilometres north-east of Moscow and has only slightly more than 200,000 inhabitants. Because of its position at the north-eastern extremity of the continent, this European territory has throughout history found itself on the fringes of the large communication networks which, in the west of Eurasia, are found essentially between the lower latitudes of Europe, the Mediterranean and the Near East. Historically, then, this backwater of the European continent has been subjected more to Nordic and Siberian influences. We are here squarely in what linguists, sociologists and anthropologists call 'peripheral zones': regions a long way away from major networks which retain ancient language structures, but also certain cultural and technical traditions, that have long disappeared elsewhere. In the archaeological collections of this polar region there are craft objects from the Middle Ages that still resemble certain prehistoric artefacts. There are magnificent barbed harpoons made of bone dating from later than the twelfth century which evoke in their forms and their artisanal methods procedures that have been abandoned for at least four millennia elsewhere in Europe, where they were quickly supplanted by technologies using metal. Similar bone harpoons are found from the same era across the whole of Siberia and are also well documented in polar Inuit populations. The French polar explorers Paul-Émile Victor and Jean Malaurie were able to observe them still in use in the middle of the last century.

The history, traditions, technology and knowledge of the area do not follow a universal trajectory. In the first instance they are subject to local dynamics and derive from the history of each territory. For the archaeologist, in general only technical traditions can be reconstituted and analysed in their temporal evolution. These traditions are never deployed continent-wide. They always appear to us fragmented geographically, and one can see certain technical solutions transgressing time with no regard to the technical developments which might have led, elsewhere, to the abandonment of the finest traditional crafts. This is not a matter of backwardness. We should be aware of the relativity, even the iniquity, of the notion of technological progress. Some ancestral traditions that have survived for millennia in fact preserve the memory of a high technical expertise that we no longer know how to employ. Our own technological developments involve an irreversible loss of certain skills that are among the most advanced. We end up incapable of recognizing the superiority of certain immemorial techniques which exceed on every score our own technical knowhow. Only those who are either cynical or naive concerning the position of our societies in the architecture of all the human societies that have existed on earth would smile at such statements. Claude Lévi-Strauss, in his brilliant book *Tristes tropiques*, came to the same conclusion in the 1950s, referring precisely to the example of polar communities' techniques for protecting themselves against the cold:

only during recent years have we discovered the material and physiological principles underlying Eskimo dress and the form of Eskimo houses, and how these principles, of which we were unaware, allow them to live in harsh climatic conditions, to which they are adapted neither by use nor by anything exceptional in their constitutions. So true is this that it also enables us to understand why the so-called improvement that explorers introduced into Eskimo dress proved to be less than useless, and in fact produced the opposite result to what was intended.

The native solution was perfect: we could only realize this once we had grasped the theory on which it was based.*

There are many similar examples covering the huge range of knowledge that has been lost over time and that we are rediscovering only in dribs and drabs in different technical fields: from the anti-seismic construction methods of the *machiya* in Japan to high-yield techniques in agriculture via the knowledge and very precise use of molecules produced by certain plants, millennial knowledge that is now held only by some bearers of these ancient memories and is the envy of the big pharmaceutical industry.

All these thoughts on the structure of traditional technologies, their evolution and replacement were going through my mind as I opened that drawer for the first time in February 2007. In it was a collection of several hundred carved stone tools. But carved following the old artisanal traditions of the Neanderthals that I knew so well. Drawer after drawer, I went through the collections from the Byzovaya deposit, a site on the Arctic Circle reliably dated by multiple laboratories to . . . 28,000 years.

Twenty-eight thousand years? That places us a full fourteen millennia after the time when Neanderthal artisanal traditions disappeared from the whole of Eurasia. Piece after piece, I recognized the technologies, the tools, the style, the knowhow, diagnostics of a traditional technology that had disappeared everywhere else with the eviction of the Neanderthals. Byzovaya is more or less the same age as Yana RHS. But Yana is a long way away – occupying the same polar climes but more than 2,000 kilometres distant, in the east of Siberia. The two sites share a common chronology, their boreal location and a

* Claude Lévi-Strauss, *Tristes tropiques*, Penguin, 2011.

great interest in hunting mammoths. But at Byzovaya the thousands – rather, the tens of thousands – of sophisticated objects made from animal matter found at Yana – pendants, ornaments, bowls, javelins, bracelets – are simply absent. Where the mammoths at Yana RHS were hunted almost exclusively for their straight tusks and not for their meat, at the same time, 2,000 kilometres away, on the other side of the polar Urals, the men of Byzovaya displayed no interest whatsoever in the transformation of ivory, reindeer antlers or bones. A study of mammoth remains found that the pachyderms had probably been hunted. Analysis of their bones revealed very clear traces of their exploitation: opening of the rib cage, filleting, the breaking of ribs and other acts of butchery which showed that the relationship to these remarkable herbivores was based here on completely different objectives. In Byzovaya, no interest in ivory tusks, but a great interest in the animal's flesh. Two completely different logics, which immediately distinguishes Byzovaya from its Siberian sister Yana. The stone tools themselves were easy to classify in the wider Mousterian family. The Mousterian is an old technical tradition of producing stone tools that was practised by the Neanderthals in Europe for hundreds of thousands of years.

What were these collections trying to tell us? At this age Europe as well as the Siberian far north were occupied by different *sapiens* populations, who were the legatees of distinct technical and cultural traditions. But these traditions had in common a singular relationship with the animal world, the search for their hard matter, bone and ivory, which were systematically transformed into tools and invested with very strong symbolism. Decorations, jewels, representations are all the exclusive prerogatives of *Sapiens*. And there is a symbolic dimension to these tools of bone, reindeer antler or ivory. Animal matter was used as adornment: the animal provided both covering and decoration for humans. Remarkably, when an ivory javelin was used in hunting, the animal was killed . . . by the animal itself. This brings to mind the traditions of modern Siberians, Inuits and Plains Indians, who must protect themselves from the spirit of the dead animal, defend themselves from the ghost that might come to torment them. They talk to its spirit, even before going to hunt it, they take its side, like a friend. They speak to the animal, caress it, after having killed it. Don't get this wrong: they are not apologizing, they are thanking the animal. They thank it and explain that its flesh will feed their children, that its skin will keep their old people warm; they fear the magic power of its spirit, and so they associate themselves with it, connect the animal to the human community, to which it will henceforth belong. And by using an animal to kill an animal, they integrate it in the hunt of its fellow creatures.

Humans make themselves a part of their natural environment, with which they form a continuity, with which they intimately associate themselves. Animist structures. Shamanistic structures. In these gestures, in these actions by the *Sapiens* of the European Palaeolithic, we can see the logic of all modern hunter-gatherers, in the way they interact with the natural world to this day. This millennial logic of *Homo sapiens* in Europe is totally absent at Byzovaya. Instead, we find traditional Neanderthal handicrafts of a totally different order, forms of tradition that should no longer have existed in such recent periods.

The question of the persistence of the Neanderthal, or perhaps simply the persistence of its traditions, its handicraft, beyond the forty-second millennium has been much debated in the scientific community over the last twenty years. In various places across the continent we have uncovered Neanderthal remains or Mousterian tools, the famous material culture of the Neanderthals attributed to slightly more recent phases, perhaps only thirty-five millennia ago, perhaps here and there as late as 30,000 years ago. And yet, we have gradually come to realize that these Neanderthal bones, in Belgium, Croatia, Spain, the Caucasus, were probably not so recent, and problems with dating may have affected their chronological attribution. Why such discrepancies? Neanderthal bone remains are a very rare and precious archaeological commodity. They were commonly uncovered in older archaeological excavations, when research was based on digging in caves with a pickaxe, displacing over the course of few decades thousands of cubic metres of sediment through an almost industrial exploitation of the great caves of Europe. Bones were turned up: sometimes complete Neanderthal teeth and skulls. Nowadays, across the continent, they form our main archives on these extinct people. And their archaeological context is as precise as can be defined by a pickaxe. Some of the caves had been regularly occupied by Neanderthals for more than a hundred millennia, and over that time hunters left tens of thousands of tools, tens of thousands of bones, evidence of their day-to-day activities in these locations. And the huge archaeological collections have got mixed up, no longer distinguishing remains abandoned by the Neanderthals forty-two millennia ago from those abandoned 120,000 years ago. All of it is Neanderthal, but bear in mind that the Neanderthal who died in this cave 42,000 years ago is much nearer to us in time than the Neanderthal buried in the cave 120,000 years ago. Much closer to us in time . . . But not in other ways.

These bodies, these objects, now orphaned from their precise contexts, no longer have much meaning for the archaeologist, who is confronted with an unsolvable puzzle, or rather an Olympic-size swimming pool in which the pieces of millions of jigsaws are all mixed up but all look alike. Good luck with that. These Neanderthal bones, so beautiful, so rare, were stored in boxes, drawers and bags, were glued, solidified with resins, handled by the sweaty hands of generations of learned admirers who came from all over the world to study them. And after 100, or 150, or 170 years of regular handling, at the turn of the twenty-first century, eminent researchers came along to scrape some powder off these bones to extract some molecular information from it: their age by carbon-14 dating, their genetic history – from the 1990s onwards – their isotopes or the fossilized proteins they contained. Science, real, hard science, of course, but hard science with feet of clay. The foundations of the building are worryingly fragile. This whole history that we researchers have built and continue to build is founded to a large extent on analyses – molecular analyses – of a very high resolution but based on scattered jigsaw pieces without any image of what the overall jigsaw should look like. Without any serious archaeological context, the age of these human remains is largely determined by totally decontextualized carbon-14 datings. There is nothing more fragile than a molecular analysis without a robust context. The right-hand side of your bone might be dated to forty millennia and the left to less than 20,000 years. Make no mistake about it, a huge number of conclusions derived from these molecular analyses are debatable. These conclusions, these hypotheses cannot rely upon current archaeological techniques, which are the only ones that enable us, in the first instance, to reassemble the puzzle before analysing the image. Consider the fact that, in the Mandrin cave in the middle of the Rhône Valley, which I am currently excavating, we have had to dig using brushes and small bamboos two to four months a year for thirty years just to get down 60 centimetres into the archaeological sediments of the cave. There are now almost no more Neanderthal remains. The last time a body that was anywhere near complete was discovered in France was in 1979, forty-four years ago. And its archaeological context, its age and the culture this person was part of are uncertain and the object of intense debates that might never reach a conclusion.

Given such limitations, scientific proof is fragile and dependent on the reliability of biomolecular analyses. It's a leap of faith. Our key archives derive from these old excavations. And our archaeological or molecular documentation, our analyses and conclusions on these populations are remarkably vulnerable. Scientific texts are soon out of date, which, in itself, is a sign that the discipline is in good health, but it is not much help in defining what these astonishing Neanderthal populations were really like and how they lived, and even less when and why they became extinct.

So . . . did the Neanderthal survive beyond the forty-second millennium, as a number of recent scientific analyses have suggested? The proposition seems increasingly shaky. The most recently discovered Neanderthal remains seem to be older than the others. Each new analysis produces an earlier date, the extinction progressively acquiring a more ancient chronology. A few sites still seem to counter these critical analyses of the archaeological documentation on the last Neanderthal societies. But debate is raging in the scientific community about whether Neanderthal populations could have survived longer at the southern extremity of Europe, in the south of the Iberian peninsula. There are no Neanderthal human remains here, just collections of archaeological evidence attributed to Neanderthal cultures by an analysis of the technologies used to make their stone tools. These Mousterian technologies, which are associated exclusively with the Neanderthals in

Europe, were still in use there as late as thirty millennia ago. The scientific discussions concern the quality of the carbon-14 datings. The collections from the southernmost tip of Europe are affected by the warm Mediterranean climate, and bones, one of the main materials used in carbon dating, frequently yield aberrant results which suggest chronologies significantly more recent than the actual age of the objects. Small degrees of pollution can lead to considerable errors. A tiny 1 per cent of recent carbon can make a bone seem seven millennia younger. The problems of conservation and pollution of bones are now well known. One way to deal with them has been to measure not the bones themselves but the age of the carbon preserved in the archaeological layers around them. The results are in fact deceptive. The large spread of readings obtained can sometimes cause ages to double. Here, the problem is less likely to be the method itself than the way the small traces of carbon shift through archaeological layers. These very small, light objects have a tendency to drift across millennia in archae-ological soil due to the action of micro-roots or the flow of water. These coals are undoubtedly recent, but did they really come from the fires Neanderthals lit in their caves? Or are they simply intruders, tiny vagabond objects? It is currently impos-sible to know. So we will continue to wonder whether ancient aboriginal populations survived in the extreme south-west of the European continent.

In Byzovaya we have the mirror image of these questions of Neanderthal survival. At the north-east tip of the European continent, in archaeological layers that are unquestionably twenty-eight millennia old, we have discovered hundreds of carved-stone objects, purely Mousterian, so purely Neander-thal according to our historical understanding of technological developments in Europe, abandoned on the Arctic Circle and quite unlike objects found elsewhere. As we have seen already, Byzovaya is one of only three recognized ancient sites in the polar regions. But this time, the Arctic climate has had exactly the opposite impact to that of the Iberian peninsula when it comes to the preservation of bones dated using the carbon-14 method. More than forty datings were made on the bones of Byzovaya. Here, there were no discrepancies, no range of readings spanning millennia, as in Spain, but rather a set of precise, tightly grouped dates. Here all lights are on green. The Byzovaya bones were excellently preserved and especially conducive to carbon-14 analysis. They had spent most of their existence in frozen ground, which perfectly preserved their collagen. Numerous measurements all pointed to a single phase of Palaeolithic occupation, centred around 28,000 years ago. The data are robust, very robust. Professor Tom Higham of Oxford University, one of the world's top experts on carbon-14 dating, said that Byzovaya was one of the best-dated Mousterian sites in the world. This time, it wasn't the wandering micro-carbons that were dated, but whole mammoth carcasses, anatomically connected, whose bones bore the marks of the Mousterian flint tools which were used to cut them up. In 2007, looking in these open drawers in the Komi Republic, a totally unexpected historical scenario began to take shape in my mind. How come there was a Mousterian presence on the Arctic Circle, fourteen millennia after it had disappeared from the rest of the planet?

Nordic Transgressions, East and West

To attempt to address this boreal enigma, we had to return to the banks of the great river Pechora on the Arctic Circle. So, along with a group of French and Russian researchers, I mounted a mission supported by the French foreign ministry to explore this question of the colonization of the European polar regions. I was happy to be working in the polar taiga: this European wilderness looks like some of the finest regions of Canada's far north. Vast and beautiful.

To get to Byzovaya you catch a train from St Petersburg all the way to the great north-east. The journey takes a few days, and you pass through the huge virgin forests of the European far north. The train has to travel at a cruising speed of around sixty kilometres an hour, a nostalgic throwback to the Orient Express or the Pacific Railroad, the first transcontinental steam train to the American West. Each carriage, a liberal mix of iron, wood, aluminium and lace curtains, is a self-contained shelter providing water, heating and enough food for several days in case of a breakdown in winter. Up here, winter lasts for virtually eight months of the year. You have to bear in mind that at the height of this long cold season, at -35 °C, and without any feasible form of transport other than this rail track, a breakdown or power outage can leave you stranded for several hours. Or several days. The autonomy of the carriages is not merely an optional extra. At each end of the carriage a coal boiler provides steady hot water for the heating system and keeps the lovely silver samovar permanently topped up.

As the earth is a sphere, the three continents of the northern hemisphere join up at their northernmost edges. Patrick Plumet, the eminent specialist in prehistoric societies of the far north, noted that all the peoples of the Americas bear the trace of their boreal origins, fossilized in some of their traditions, a distant reminder of their polar origins and their original crossing of the Bering Strait. Travelling north inexorably means drawing together Europe, Asia and America. In the far north, all borders inevitably meet. Dealing with the European far north is also dealing with all these continental borders.

Let us return to our Orient Express with its slowly moving horizons. Tea, coffee on tap and heating in all parts – simple luxury, old-fashioned and delicious. Children quickly appropriate the communal spaces, running up and down the corridors from one carriage to another; life is lived here without a filter of reserve. At every stop old people rap on the windows; they have tasty delicacies to share: blueberries, mushrooms, dried fish and, depending on the season, vanilla ice cream. During such stops – we never know exactly how long they will last – we are treated to exquisite displays of young Russians climbing down to the platform, cigarettes dangling carelessly from their mouths, in flip-flops and shorts, just to spend a few moments on terra firma. You have to make yourself comfortable on these journeys, when your destination is such a distant prospect. 'Shorts and flip-flops' is pretty much the dress code for the stops: even in February, when the temperature is -25 °C, you pop out bare-legged to enjoy a very frozen vanilla cornet on the platform. It is a mixture of the body's adaptation to extreme temperatures and the sociological expression of a frank and open virility. A treat for the eyes and the mind.

This slow journey seems endless, but it is grand and wild, apart from the occasional industrial stretch. Although there is nothing but taiga for hour after hour, we pass an infinite number of chimneys in ageing concrete and rusted metal stretching from one horizon to another. Huge factories are spewing out their poisonous fumes in all the colours of the rainbow. Along the sunken lanes, the standing water also has a distinctive colour, tending rather to fluorescent green, which adds its own original touch to this *Mad Max* world. Nausea. Why can't this damn train go faster? Our caravan finally arrives at Pechora.

There, straddling the Arctic Circle, we find a city of 40,000 souls, built in 1940, with its railway station and its port, plonked down next to the river of the same name. It has all the elegance you'd expect of a Soviet labour camp, albeit a polar one, which received hundreds of thousands of prisoners up until 1959. It was these new serfs who built their camp with their own hands. The city is snowed in for around 200 days a year. For your comfort you can expect little more than two months a year without frost. There is rust everywhere, like some comic-book vision of the end of the world. But gloomy feelings are offset by the warm welcome of the Pechorians, a Russian warmth which is generally only superficially hidden beneath a rigid facial mask. The open kindness of the people lends this morose environment an almost enviable quality. After a visit to the local museum and an inspection of their collection, we kit ourselves out in the local style at the warehouses of the merchants of hunting and fishing gear. Parkas for rain, thigh-length boots and above all netted anti-*mochka* hats. *Mochka* is a sort of boreal cross between a mosquito and a fly. A kind of mini mosquito which, instead of stinging, bites and tears at your flesh. The problem is that in summer your body is covered with them, and they are so small that they manage to slip through the holes in the nets suspended from our hats.

A few hours' drive from our former gulag, we reach Byzovaya, on a bend of this huge boreal river. Here, by contrast, everything is natural. And everything is naturally grand. Right down to the *mochka*, which oblige us, as soon as we stop, to burn old stumps of damp fir and birch to make smoke to drive them away so that we have a safe refuge in the taiga during the time of our research. The camp is in a clearing very close to the dig. The view is stunning: we overlook the great river and forest expanse stretches out before us to the massive mountains of the polar Urals on the horizon. We cross paths with an old man off to try his luck fishing, his cane rod over his shoulder.

'Ты француз?' ['Are you French?']

'Yes.'

Cue huge peals of toothless laughter.

'Даже Наполеон не дошёл так далеко!' ['Even Napoleon didn't get this far!']

A reference to 1812, the retreat from Russia, when Napoleon recounted that his fallen horses could not get up again because of the cold and men burned to death, huddled too close to the fire. But now the only quadrupeds that interested us were the mammoths.

The Polar Refuge of the Last Neanderthals?

The Byzovaya site is on the banks of the river, buried under metres of silt and sand deposited by the wind during the last ice age. Thanks to the slope and the erosion of riverbanks, the archaeological layer is fairly easily accessible by means of a series of trenches that we dig to reveal the vestiges of human settlements. As soon as we reach the archaeological layers we find fine mammoth remains together with carved stone tools. The site contains an impressive accumulation of fossilized mammoth bones. There are elements clumped together, buried deep. The site does not seem to have been at all altered by time, and the remains are undoubtedly in their primary position. Analysis of the bones shows that these people were here principally to take advantage of the large pachyderms. Prospecting along the river, I pick up stones from the main rocks used by the prehistoric hunters. On the way to Pechora I bought from an artisan some elk antlers found in the surrounding taiga. With these few antlers and a selection of stones gathered from the river I began to fashion the same tools that the prehistoric people had used to bone their mammoths. The rocks lent themselves well to it, and I was able to reproduce the main types of Palaeolithic tools that I had analysed at Syktyvkar. It was obvious that the artisans had great mastery of the materials, the essential sources of their craft. It seemed equally obvious that the stones gathered from the river lent themselves to a wide range of very different tools, and that our mysterious people of Byzovaya had the means to manufacture here, if they so wished, objects of many different shapes. At the end of the mission we spent a few days studying the collections in the museum at Pechora. As usual, we had a warm welcome, and the director of the museum brought me a hot tea and . . . a plastic bag. It contained a superb carved flint point polished smooth by the effects of water. A 'Levallois' point characteristic of Neanderthal technology. It isn't from Byzovaya – a child found it in Pechora itself

Before we caught our train back to more southerly climes, the director of the museum took me to the place where the discovery was made, where despite a day spent prospecting I myself discovered nothing but the dilapidated landscape of the old Soviet port. That stone point remained as an isolated item on a map. Just one more boreal enigma.

A few weeks later, back in St Petersburg, I analysed a large corpus of carved stones contemporary with Byzovaya but originating from excavations situated in more southerly zones, from the sub-Arctic regions to the great Russian plain which extends from the Caspian Sea to the Black Sea. One advantage of Soviet archaeology is that the archaeological collections are remarkably centralized, so from the comfort of my seat I could travel all across central and eastern Europe as far as the borders of Iran simply by opening, one after another, the wide wooden drawers ranged beneath the magnificent painted ceilings of a former palace of the tsars. With a generosity that one can only envy, my Russian colleagues gave me access to the whole of their archaeological treasures. In this hushed atmosphere amidst the old-fashioned charm of the post-Soviet furnishings, I opened the drawers . . . I examined a large number of carved stone objects in archaeological collections from more southerly regions of Europe and found myself face to face with very classical artefacts from the European Palaeolithic. These were artefacts from societies contemporary with Byzovaya. To better understand the traditions of these prehistoric societies, I also included in my study some significantly more ancient or more recent sites, creating a vast temporal panorama encompassing the evolution of the traditions of the Palaeolithic populations of eastern Europe.

The conclusion is surprising, but indisputable. The artefacts of Byzovaya have no obvious connection with the neighbouring regions to the south as far as the barriers of the Black Sea, the Caucasus and the Caspian Sea, whose collections I analysed specifically. We were here on the European Arctic Circle looking at technically very homogeneous collections, archaeologically well located and accurately dated to 28,500 years ago. These technologies were very familiar to me and were, without a shadow of a doubt, strictly Mousterian. In Europe, the Mousterian has never been connected with any other humanity than the Neanderthal. Human remains are few and far between, and it is almost exclusively Mousterian objects that allow us to identify a Neanderthal presence in a site. But we are on the Arctic Circle, more than 1,000 kilometres north of the most Nordic Neanderthal sites. There is a chronological leap here that is enough to induce vertigo. These traces are dated a good fourteen millennia after the presumed disappearance of Neanderthals from the planet.

So who exactly were the people of Byzovaya?

It was in response to this fundamental question that in 2011 my colleagues and I published an influential article entitled 'Late Mousterian Persistence Near the Arctic Circle' in the prestigious journal *Science*, opening up a debate on this remarkable and unsolved enigma. We were able to draw three fundamental conclusions: the artefacts of Byzovaya are strictly Mousterian; their age is unquestionably 28,500 years; and in Europe it is exclusively the Neanderthals who are associated with these artisanal traditions.

Of course, we would like to have found human bones in order to go further.

With only three sites on earth to inform us about the colonization of high polar latitudes, all we can do is open doors and construct hypotheses: any conclusion can only be a fragile one. No one can be sure who these enigmatic populations really were. But just imagine for a moment that the last bearers of this Mousterian culture were *Homo sapiens* . . . This discovery would be just as remarkable, just as unsettling in its implications for our understanding of the distant boreal civilizations of the Palaeolithic.

Across Africa and as far as the Near East, but in much more ancient times, *Homo sapiens* were effectively inheritors of technological traditions very similar to those seen among the Neanderthals. But we also know that at the time of their colonization of the high latitudes of Europe and Eurasia, generally speaking, the *sapiens* populations had traditions that were distinct from Neanderthal traditions. Their remarkably modern new technologies were based on creating standardized stone tools; their weapons, bows and spears, relied on the systematic use of mechanical propulsion. Art and very rich jewellery in bone or ivory were generic markers of *sapiens* societies at the time of their colonization of Eurasia. In the southern regions of the Mediterranean Levant, where *Sapiens* was known to have lived well before the hundredth millennium, such new practices were in evidence in very ancient times, probably before the fiftieth millennium, rapidly supplanting the old Mousterian traditions at these latitudes. It was only when equipped with their new artefacts, their new practices, that *Sapiens* would rapidly conquer all the biotypes of the Eurasian supercontinent. This lightning-fast and all-pervasive colonization took over the whole of the traditional territories of the aboriginal Neanderthals with remarkable speed. And these innovations had spread throughout the *sapiens* societies of Eurasia more than twenty millennia before the polar occupations of Byzovaya. From the fiftieth millennium, *Sapiens* had thus abandoned the ancient technological traditions of the Mousterian, and these technological traditions would survive only in continental Neanderthal populations until their biological extinction a few millennia later.

We have also seen that, in the Siberian polar latitudes, the Yana RHS site was contemporaneous with that of Byzovaya and that archaeological research uncovered two *sapiens* child's teeth, the DNA of which reveals genetic exchanges with the only Neanderthal populations. But Yana is 2,000 kilometres as the crow flies from Byzovaya and has yielded an abundant diagnostic array of modern traditions in its quasi-industrial exploitation of mammoth ivories and its thousands of pieces of jewellery or bone sewing needles.

Another discovery, unfortunately much further south, at least informs us about a more ancient population occupying Siberia – to date the oldest known Siberian population. In this case it is an isolated object. A bone. The fragment of a femur, to be precise. It was discovered in 2008 by Nikolai Peristov, a Russian artist who creates jewels, necklaces and sculptures from the ancient material of ivory. Peristov found the bone on the banks of the river Irtysh in the west of Siberia, and he brought it to the local police station, where it was identified as human. Its blackened patina, its weight and its mineralization left doubt as to its age. The police logged this as a bone of great age, as its texture closely resembled that of the Palaeolithic bones harvested by Peristov. It was then subjected to datings and genetic analyses which showed that – astonishingly – this human leg bone was . . . 45,000 years old. And its DNA was perfectly preserved, having spent a large part of its existence in frozen Siberian soil. The bone of Ust'-Ishim yielded the most ancient *Homo sapiens* DNA ever obtained on earth. We are in western Siberia, east of the Ural mountains, in the vast expanses between India and the Taymyr peninsula, where the carcass of that mammoth killed and butchered with stone tools was found. However, our Ust'-Ishim man was a good three millennia short of the age of the Taymyr hunters and 1,500 kilometres in a direct line south of the polar regions. Wrong time, wrong place, then: the jury is still out.

But these genetic data give us leads that are at least interesting. The population of Ust'-Ishim is totally distinct from that of the 'Ancient Siberians of the North' identified far to the east, at Yana, but it too, like Yana, bears witness to a human branch that became extinct without any direct descendants in modern populations. Even more interestingly, just like at Yana, Ust'-Ishim reveals no genetic contact with the Denisovan populations identified in the neighbouring Altay mountains. And, as at Yana, this *sapiens* population without descendants counts the Neanderthals among its ancestors. Genetic analyses allow us to calculate the age of this encounter between the Neanderthal and *Homo sapiens*, which took place eighty-four generations – that is, between 1,500 and 2,000 years – before the birth of Ust'-Ishim man. The encounter therefore took place 46,000 to 47,000 years ago. Its Neanderthal baggage is near in time and could correspond to the very first phases of the polar populations, which are known to us through the remains of a mammoth and a wolf. In our two fossil *sapiens* populations, the Ancient Siberians of the North and the one at Ust'-Ishim, Neanderthal has left its imprint in time, while Denisova, the Asiatic cousin of the Neanderthal, is absent from the family reunion . . . These clues appear compatible with a point of encounter with the Neanderthals much further west and potentially also very much further north, since it was situated outside the more southerly regions occupied by the Denisovans.

But the first boreal populations, the ones we have traced back forty-eight millennia 600 kilometres north of the Arctic Circle, are a complete mystery to us. The analysis of vestiges of their game is unambiguous. There were humans there. We know that human populations lived in the ultra-boreal regions very far back in time, at a time when, as far as we know, the middle latitudes of Eurasia were occupied by now-extinct populations, the Neanderthals and the Denisovans, even before the Neanderthal–*Sapiens* encounter evidenced by the genes of Ust'-Ishim man. In these high latitudes distant human presences continue to be inaccessible. We know that humans were there, chasing mammoths and wolves, but we haven't found a single artisanal object, and we know nothing of their traditions, their organizations or the time when they arrived so far north of the Arctic Circle, at the edge of the known world. They remain an enigma.

Was the Neanderthal a polar creature? No one knows, but

our research tests the limits of our knowledge of these populations, and our polar explorations raise doubts. The doubts that
dog researchers at every step as they try to understand and
come face to face with this extinct humanity.

From the People of the Mammoth to the
People of the Whale

We can only wonder what became of these amazing boreal
civilizations.

It is possible that they abandoned the very high latitudes when
the great cold struck the planet a few millennia after the polar
settlements of Yana and Byzovaya – a glacial episode that would
last for seven to eight millennia. But there are no data to support
this, and for the moment we are completely in the dark about
these old populations of the far north. The global rewarming
that began 11,700 years ago brought about major environmental
changes in the boreal zones, and areas once teeming with game
became radically impoverished with the disappearance not just
of the mammoth, but also of horses and bison, in the most
northerly regions. These polar expanses turned to marshland,
unsuitable for the development of large herds of herbivores.
The response of the populations to their new Arctic environments was impressive. Several millennia after this rewarming,
archaeologists discovered hyperboreal camps on Zokhov
Island, a small island of 77 square kilometres, the exact same
size as Guernsey, a flake of confetti in the frozen Arctic Ocean,
but one located 1,000 kilometres north of the Arctic Circle.

Extraordinary hunting activity has been documented in camps
which are older than 9,000 years. These Zhokov groups organized their food economy around the systematic exploitation of

polar bears, the main land-based large mammals in these very high boreal latitudes. Polar bears are formidable predators, but they were exploited at Zhukov for their meat, hunted with the aid of lances and javelins, which were stuck in their neck or at the base of their head. This was a dangerous undertaking, and the polar bear on its own could not make up for the vast resources formerly available during the ice age on the mammoth steppes. More recent archaeological data indicate that the populations would no longer look to the polar bear itself but to that animals' own prey, marine creatures. While hunting these resources would be less perilous, it required a complete reorganization of polar societies, of their weapons, their artefacts, their traditional customs. The exploitation of marine resources – fish and mammals – draws an invisible millennial line between the peoples of the mammoth and the peoples of the whale. The peoples of the whale were coastal Inuit populations, which organized themselves – indeed still do – around the resources of the far north and Greenland. There are very few ancient sites along this boreal Chinese wall, and after the passing of millennia it is not easy to join the dots that would enable us to write the troubling and fascinating history of the first peoples of the far north.

3.
Cannibals in the Forest

Neanderthals are an enigma, whether it's to do with their true nature or the unknowable reality of their polar existence. The nature and significance of their actions more than 100 millennia ago at our latitudes are barely comprehensible to us any more. It is as if mystery was the defining feature of this extinct humanity. More than 100 millennia ago, in the lush primary forests, so different from the polar steppes that we have just crossed, the discovery of deliberately broken, chopped and dismembered human bones suggests that some of these peoples of the forest were cannibals. Really? Cannibals?

Eat Up, Everyone

'Anthropophagy seems to be one of those customs it is easier to acquire than to give up.'

No, not a quote from Hannibal Lecter in *The Silence of the Lambs*, but the words of the ethnologist Hélène Clastres, writing in 1968 on the widespread practices of cannibalism in the societies of South America.

And maybe a somewhat startling idea at first sight.

Anthropophagy is the act of eating a human body. In whole or in part. More precisely, it is the act of consuming the remains of an individual, for it is not necessarily just flesh that is ingested, but other parts too: hair or bones, for example. It was bones that were commonly eaten by the peoples of the tropical forests evoked by Clastres; flesh or bone often related to distinct ritual practices. Cannibalism revolts us, fascinates us, raises questions. But this surprising practice is universal; it can be found throughout the ages, often where we least expect it. Perhaps even you practise cannibalism without realizing it. Please read on a bit further before you take offence, all will be revealed.

Far from being a tradition alien to our western world, the eating of human matter is deeply ingrained in our history. It was commonly practised in Europe from the Palaeolithic to the Iron Age virtually without interruption. Archaeological data clearly indicate that cannibalism was part and parcel of numerous societies up to our Celtic ancestors. In the first century CE, Pliny the Elder described such practices and the ritual significance given them by certain Celtic populations. This is no mere casual accusation on the part of the Roman naturalist, no politically motivated attempt to paint the ancient Celts as barbarians. Contemporary archaeology shows us without a shadow of a doubt that such practices were widespread in Iron Age societies. The image of Asterix the Gaul takes rather a knock, though imagine how the final banquet scene in Goscinny's books might have played out . . . The theme of cannibalism also permeates the imaginative landscape of our Middle Ages, and accusations of cannibalism still recur in the stories and fantasies of our societies. The spell books that occur throughout our history, from the Middle Ages to modern times, and which enjoyed their widest diffusion in the nineteenth century, present many recipes requiring the consumption of flesh, or human blood. In the seventeenth century, *Little Albert*, one of the commonest spell books, advises the following: 'Draw some of your blood on a Friday in spring, leave it to dry in a small pot with the two balls of a hare and the liver of a dove, reduce everything to a fine powder, and make the person on whom you have designs swallow it, about half a gram, and if the effect is not achieved the first time, repeat up to three times and you will be loved.' Love and cannibalism, then, are linked.

Such practices were violently suppressed by the Catholic Church through the enforcement of the Inquisition. Catholicism itself was not backward in the matter of ritualized cannibal practices, since they in fact lie at the very heart of the religion: 'This is my body.' Then, taking a cup and giving thanks, the priest says: 'This is my blood.' These phrases offer little room for interpretation; indeed the Council of Trent, in 1551, asserted the literal reality of transubstantiation. The host and the wine are not symbols of the body and blood of Christ but contain the 'real presence' of this body and this blood, which are then consumed. As in the *Little Albert* spell book we can see a subtle but well-defined line – an unconscious one, but quite distinct – linking love and the ingestion of the person who is loved. A little like the way we might say to our beloved: 'I could eat you up.'

Cannibalism is ubiquitous and seems constantly to escape any rigid definition. In the end, it is not just about making a steak tartare out of Grandma but speaks to us about our emotions, our relationship with love and the ways we accept death, provided that something survives within us. Eaters of humans come in many forms, and they pose questions about practices that are simultaneously the most common and the most rejected by our societies. Today, in popular culture, we still instinctively perceive cannibals as barbarians unlike ourselves, as dehumanized beings. The super-intelligent Hannibal Lecter, the werewolf, the vampire who regenerates himself by drinking human blood: all are inhuman and yet, at the same time, superhuman. Cannibalism shows us the unconscious, repressed path of something which transcends us, is greater than us . . . 'Jesus said to them, "Most assuredly, I say to you, unless you eat the flesh of the Son of Man and drink His blood, you have no life in you"' (John 6: 53–8).

From the middle of the nineteenth century, ethnologists and prehistorians began to understand that there was an unexpected cultural depth to these practices. So the idea emerged in the nineteenth century that they would have been too complex to be exercised by the Neanderthals, then known as Mousterian Man:

> The opinion [is] that this barbarous custom presupposes a certain degree of civilization and metaphysical ideas on the distinction between the soul and the body. The savage eats his enemy to assimilate both his qualities and his courage and to double his own energy; also, convinced that he has absorbed the whole personality of the valiant man, he often abandons his own name to adopt that of his victim. Such an assembly of concepts has little relation with what we know about the customs and the intellectual level of Mousterian Man. If the theory of M. Carl Vogt is well founded, these people would have been cannibals only by chance, and, however strange the assertion may appear, they would not have been civilized enough to be anthropophagous.

Not civilized enough to be anthropophagous! These ideas were expressed in 1873 by the French aristocrat Baron Lubac following his archaeological research in the Néron cave, in Ardèche, where he discovered burned human bones in the hearths, mixed in with the bones of game hunted by the Neanderthals.

Cannibalism without Appetite

The extraordinary archaeological collections from the Néron cave were scattered far and wide, and there is no study documenting the question of the treatment of bodies by the Neanderthals who resided there. One hundred and twenty-six years later, in 1999, in a small cavity situated 20 metres or so below the Néron cave, evidence was turned up of the practice of cannibalism in some Neanderthal societies. But what sort of cannibalism are we talking about? I was one of the authors of the study published in *Science*. I first encountered this fascinating cavity at the age of twenty and spent some of the best years of my training in prehistoric archaeology digging there two months a year over the course of six successive years. By the end I had spent the equivalent of a whole year in the rock cavity. We worked there in a small team, and it was my first real contact with the Neanderthal people.

The cavity was like a sort of natural well scoured out of the rock. Descent was via wooden ladders, and at the bottom, trapped in a small ring of stone about 20 metres across, we set foot on archaeological layers going back 120 millennia. An interesting and little understood era when the earth's climate switched violently from a glacial period to a temperate period. Very temperate. Much warmer than now. The great grassy steppes of the boreal climates were rapidly replaced by woodland; open landscapes as in present-day Mongolia gave way to very rich, dense forests. Vast primary forests that seemed to go on for ever. Forests in which no tree was ever cut down, no road was ever opened up. The oceans were on average 2 °C warmer than they are today, and significant marine inundations were recorded. The melting of huge continental glaciers generated a rise in sea levels of several dozen metres; sea level was thus 6 to 9 metres higher than it is today. Landscapes changed very rapidly. Not just vegetation, but the whole biotope lurched into a whole new equilibrium. The great fauna – the horses, mammoths, reindeer, bison – which were previously dominant were quickly replaced by forest species, in particular cervid, roe deer and the amazing megaceros, a giant deer whose antlers could measure 3.5 metres across. The length of a car. These immense forests also hosted a very broad spectrum of carnivores: hyenas, lions, panthers, wolves. Continental archaeological records reveal a veritable explosion of biodiversity in response to the brutal climatic fluctuation. Small species – rodents, chiroptera, amphibians, snakes – developed a remarkable diversity, four to five times higher than that during the preceding ice age.

In the cavity, long days were spent digging with small bamboo tools so as not to damage the archaeological objects when they were extracted from the soil. Sediments were washed in sieves with very fine meshes, then dry sorted in order to recover micro-bones of the smallest species, fundamental for reconstructing past environments and climates. While the aim of the research was fascinating, the day-to-day work was boring, and often not very edifying. The cavity yielded up hardly any flints and only the occasional piece of bone. The archaeological layers themselves had been subjected to rather unprofessional digs by a local weekend hobbyist using a pickaxe. Since the cavity delivered little in the way of archaeological remains, its importance wasn't recognized soon enough. In the 1970s, the archaeological authorities believed that this small hole in the rock contained only displaced remains, which had perhaps slid down the slope from the vast cave above, the Néron, that cathedral of the Palaeolithic which was listed as a historical monument in 1965. It is likely, then, that the archaeological authorities allowed Pierre, our amateur, to access this small hole as a way of harnessing his enthusiasm to extract some data from one of the innumerable cavities that pockmark the hills of the Ardèche. At first sight this was a reasonable strategy.

It might well have paid off, and Pierre did his best, with his mixture of passion, energy and limited knowledge. He dug deep into the sedimentary archive of the cave before he ran out of steam. His was a vocation of former times. Dig with a sense of poetry and descend into the entrails of the earth. There were very few archaeological remains relative to the amount of earth excavated, but the collections were lovingly stored away by Pierre, who would cut out profiles in white polystyrene blocks to hold a piece of flint or a horse's bone or a wolf's mandible. He dug, he labelled, he dreamed, he travelled back in time. But by the 1980s, the era of the amateur archaeologist was coming to an end. The discipline had rapidly become professionalized. The archaeological authorities finally decided to visit the site and take stock of the immense work Pierre had accomplished. He had made some headway, cycling to the cave at weekends on his old green Peugeot bike with its sandwich basket. The authorities found that the smiling Pierre had dug a hole 6 metres deep, and as they descended into it they could see a rich succession of sedimentary deposits, layer after layer, going back in time. There were distinct phases in the 6-metre drop, marked by differences of colour and texture in the layers of sediment piled on top of one another like a gigantic mille-feuille. A 6-metre thick mille-feuille . . .

By analysing the cross-sections on the edges of the immense excavation, the authorities soon realized that the flints and the bones that the cavity concealed could not have come from the Néron cave a few dozen metres higher up. Over the years, Pierre had opened up a new site which recorded a succession of settlements of Neanderthal hunters. Before he came along, flints and bones had been fossilized in the cavity and they had remained sealed there since the Neanderthals had left. But when? Was it 45,000 years ago or 200,000 years ago? The collections assembled by Pierre were not able to throw light on the precise origin of the objects, and their position in the different levels of this gaping hole was at the very least a matter of chance. Bones of temperate fauna were mixed up with those of animals adapted to a polar climate. This bone and this flint, neatly arranged ten centimetres apart in their white polystyrene casing, could have been separated by as much as 100 millennia in age. And now, it is no longer possible to reconstitute the true history. The dig was stopped, but the damage had been done. It is difficult to unpick 100 millennia of jumbled-up history. It is rather as if you tried to understand the reign of Charlemagne by analysing a collection of Celtic swords, Roman tiles and Renaissance statues. This was an original civilization.

In the early 1990s, a young doctor in prehistory, Alban Defleur, tried his luck reidentifying the archaeological layers of the Néron cave, a short distance above the cavity explored by Pierre. But there, 120 years of excavation had turned this natural cathedral of the Neanderthals upside down. No less than five generations of weekend archaeologists had been digging in the huge cave since the vast archaeological operations of Count Lepic in 1870. Attempts to discover hitherto untouched archaeological layers in the Néron cave yielded only very meagre results relative to the huge effort required. It involved turning over, removing and discarding the chaos of blocks that had already been discarded and turned over dozens of times by amateur diggers in the previous twenty years. All the local kids, including grown-up kids, had visited Néron looking for flints or lion mandibles. The flints from the Néron cave are very fine and evocative of their ancient times. While he was there, the young doctor took the opportunity to do a small survey of the cave below, the Moula shelter, at the base of Pierre's abandoned excavation, at the bottom of that 6-metre-deep hole. A mere square metre. Astonishingly, after only a few dozen centimetres of digging, thirteen pieces of bone were turned up in no time at all, including three teeth and seven fragments of skull. Thirteen Neanderthal human remains. One of the finest anthropological finds in France for eleven years. The work stopped, but a survey was started up again the following year.

That was how, in 1993, in my younger days, I found myself as part of the team which was uncovering the most important collections discovered in France for a very long time. We are talking Neanderthals, huge forests and . . . cannibals. Cannibals in the forest? The team had just published an article in *Nature*: 'Cannibals among Neanderthals?' A short, half-page piece that described these various human bones. They were very fragmentary, broken up into small pieces a few centimetres in size, but analysis revealed that the fracturing was not natural and indubitably took place when the bones were still fresh, at the moment when these Neanderthals died. A close look at the surface of the bones revealed fine marks characteristic of a flint blade, marks usually left when the tool is used to separate the meat from the bone. Also, the human vestiges were mixed in with animal bones, of deer and ibex, brought back by the Neanderthal hunters, which displayed exactly the same fractures, and the same marks made by the butchering activity of the hunters. It was clear to the team, then, that something very peculiar had taken place in that cave. The animal remains found alongside the human remains pointed to a temperate climate, so the events must have unfolded more than 100 millennia ago. These bodies must have been cut up and fractured in a very distant time period, before the last ice age. But to what end? Since this first study in 1993, the team has subscribed to the theory of cannibalism, excluding the possibility that the marks could be to do with ritual actions or mortuary practices.

Why did they come to this conclusion? According to the authors of the study, ritual cannibal practices would always take great care of the bones, never breaking them. It was on this basis alone that this first article on the discoveries came to the conclusion that here was a case of subsistence cannibalism as the visible marks on the bones indicated the consumption of human flesh for purely nutritional purposes. These Neanderthals would have been chopped up and consumed to the marrow in order to extract nourishment from the parts that were rich in protein: the flesh and the marrow.

In fact, there is an incredible diversity of documented cannibalistic practices across different societies, and ethnologists have not come up with any universal rule concerning the preservation of bones. There are thousands of examples of bone breaking connected to rituals which have nothing to do with acquiring the proteins of dead people. For example, the Guayupe, Arawak natives of Colombia and Venezuela, would eat powdered bones at the behest of their gods. Here the breaking and consumption of bones is strictly for ritual purposes and has nothing to do with nutrition. This ritualization of the relationship with death enables the delicate dialectic of mourning: balancing the desire to forget and the opposing need to preserve the memory of the dead person. Such ritual practices are used on dead family members and come under the category of what ethnologists call endocannibalism. This is distinct from a large body of practices called exocannibalism, in which it is no longer the body parts of one's own kin that are eaten, but those of one's enemies.

Whether those eaten are from one's own clan or an enemy clan, the latter style of cannibalism is always highly ritualized, involving considerable theatricalized enactments of the survival of the living and their confrontation with the dangerous hold of the dead. What is being waged is the huge battle for survival of the living in the face of the dead, who insinuate themselves in a multiplicity of forms in the day-to-day life of human societies. Clastres also showed that in these South American societies cannibalism seemed to be divided into two distinct families. When the dead person is someone close to the family, the bones are eaten, but with enemies only the flesh is consumed.

The distinction between endo- and exocannibalism may be a theoretical one. In all the South American groups the dead are feared. They are distrusted. They are no longer viewed as members of society. They are about to move over to the enemy. So eating the dead becomes a way of expelling them. They are no longer considered as one of us but as an entity we must treat with suspicion, one who no longer belongs to the group. Here, the practices mingle, and the distinction between endocannibals and exocannibals becomes blurred. Who should eat the flesh? The close family? Or, conversely, the members of the tribe most distantly related to the deceased? Should one eat the flesh or the bones? Should one, at the end of the ritual, crack the skull with a bow like one does with enemies? The dead person, the relative, the father is no longer a member of the family or ethnic circle but a dangerous entity, one that should sometimes be treated as an enemy, like those prisoners whose flesh is eaten, but never the bones. In theory, endo- and exocannibalism are seen in several societies on earth and can be clearly differentiated, but in practice these distinctions can seem unsafe. The practices, which are always ritualistic, can at times resemble each other, have common elements, be interchangeable in the way they deal with the body and more generally with remains, the ultimate witnesses of death.

After 1993, the consensus was that these Neanderthal remains pointed to the actions of populations that engaged in anthropophagy for nutritional rather than ritualistic purposes. The claim was based uniquely on their deliberate fracturing of human bones. But this turned out to be a misinterpretation of ethnographic realities.

Finding the Cannibalized Bodies

Back in 1993, in the small cavity, we needed to widen the available surface area at the base of the 6-metre-deep hole. The soil had been excavated in steps, allowing a progressive descent into the deepest reaches of the cavity. The hole was a sort of inverted step pyramid, with its point at the base, making for a very restricted surface area at its deepest parts. In order to try to understand the significance of this Neanderthal bone breaking and body cutting before the hundredth millennium, we needed to expand the digging area to access and search the ancient levels to their fullest extent. So the operation had to start again at the top of the cavity, 6 metres further up, to realign the edges of Pierre's dig before reaching the floor of our cannibals. The work took six years, at two months per year, and we could only do it in small teams of a few people at a time, digging the narrow steps, balancing on old planks of wood. In total, after twelve long months, we were able to extend the deepest level to a respectable 20 square metres or so. This enabled us to open a window on to the past and begin to understand the significance of the human remains we had found.

With the right equipment we might have reached the bottom in a matter of hours, but these twelve months of work were necessary, and we could not cut corners. We had to loosen, record and gently remove all the archaeological, geological and paleontological data in the cavity, layer after layer. We had to start again from scratch in the more recent soils at the top, which had gradually become filled over the millennia. This painstaking process allowed us to understand the succession of human and climatic conditions in the cave over time. At the end of it we would finally be able to start our excavation on level 'XV', the level at which our cannibals lived. By this stage, some American researchers from the University of California at Berkeley had joined the team, and they brought their anthropological expertise in identifying human remains in the midst of animal bones. Having spent twelve months of confinement inside the cave, I knew every centimetre of the cavity like the back of my hand, and our perseverance was about to be rewarded.

After years of slim archaeological pickings, of digging balanced on suspended wooden planks to realign the edges of Pierre's gaping hole, we had finally managed to clear a good working area at the base. Our feet were now only a few centimetres from the earth trodden by our distant cannibals, and we progressed from the vertical to the horizontal in order to develop a more precise reading of the events. In moving from the vertical to the horizontal, in a way we moved from geology to ethnology. The goal of the operation would now be to slowly explore the sediments in order to find the remains left behind by Neanderthal hunters more than 100,000 years ago. Those we found were sparse but often well preserved. Animal remains appeared to have been tossed, broken and cut up, strewn around a large pile of ashes indicating a hearth area maintained by Neanderthals. We finally discovered our first human remains. Bit by bit, in almost routine fashion, 'one day, one remain', as our Californian colleagues put it. Our collection of Neanderthal vestiges rose from the thirteen remains announced in 1993 to seventy-eight bones. They came from all parts of the body, from the skull to the feet, and were jumbled together and scattered over the whole area of the site, mixed together with the remains of devoured animals and sometimes around the fireplace, though none of these bones showed any traces of charring. We published the details of these seventy-eight remains in 1999, illustrating for the first time the treatment of these bodies by Neanderthals. Analysis showed that the seventy-eight small bone fragments came from six separate individuals – two children, two adolescents and two adults. This assessment was still minimal, established from bone pairings alone. Finding six individuals in a single archaeological site with a very restricted search area was significant. And all the age groups were represented in a homogeneous fashion. Our research uncovered very few flints, and we ended up finding more human remains than stone tools. The archaeological excavation would focus on these observations. At the very least some remarkable events had taken place in this cavity.

We concluded that cannibalism in the broadest sense had been practised here, even though we couldn't exactly pin down the precise nature of the extraordinary events that had occurred in this small cavity around 100 millennia ago. The ethnologist sees the whole complex play of life and death acted out before their eyes, but the archaeologist finds only the abandoned traces, which are often very difficult to interpret. Where are the rites? Where are the actions and gestures of the living? Was this skull cracked open to eat the brain, for nutrition? So that the matter of my father will survive in me and become my own flesh? Or was it broken like in those South American populations, smashed by the members of the group, crushed, as it incarnated that part of the enemy that might return to haunt us, take vengeance on us from beyond the grave?

If I Eat You, is It Love or Hunger?

There was no further news about the cannibals of the Ardèche for another twenty years until, in 2019, the publication of an especially interesting study, one with unsettling implications. It claimed that at Moula the Neanderthals would have eaten their own dead at times of extreme famine. And these famines would have been caused by profound changes to the climate which had rapidly turned the ancient boreal biotopes into dense temperate forests. The people of the steppes, hunters of horses, would simply not have managed to adapt fully to these new environments. Now surrounded by dense forests, the large herds of herbivores that they traditionally hunted would no longer be available, and it was during these episodes of food shortage that the Neanderthals would have simply eaten their own dead. Given the incredible complexity in the ways of treating our dead shown by ethnography and history, this is a very precise explanation that points to a series of well-defined actions and excludes all other hypotheses. The authors base their reasoning on the presumed age of the site, which they locate around the 120th millennium, at the very beginning of this climatic upheaval. Their study highlights a sharp fall in the number of known sites in the temperate period in Europe – there are only five recognized sites dating from this period, and only another one in France.

The main arguments seem reasonable. The fact that we do not have many sites from the temperate period could indicate a reduction in populations in a biotope in full state of flux where the Neanderthals could no longer pursue their traditional hunting strategies. Microscopic analysis of teeth demonstrated that these Neanderthals experienced regular episodes of famine during their childhoods. And the bodies seem to have been handled in much the same way as those of the hunted animals. Indeed, the animal and human remains were mixed up together after the cutting and breaking of the bones. This cannibalism was not in any way ritualistic, then. It was neither endo- nor exocannibalism. These people were fighting for their survival in an environment in a state of change and were compelled to feed off the flesh of their dead to stay alive. At this time, the effects of climate rewarming were formidable, and the survival of these populations depended on such acts of despair.

I was never convinced by this hypothesis, and in 2020 we published a commentary in the same scientific journal, *Journal of Archaeological Science*, examining point by point the interpretation of this site where I had lived for many years. Our reply – 'Cannibals in the Forest?' – would offer a different, even opposing, interpretation by highlighting the difficulty of reaching any firm conclusion given the archaeological facts to hand.

We first of all pointed to evidence that there may be more than eighty sites in Europe attributable to the temperate period, not just five isolated sites across the continent. That seems to be an enormous discrepancy, but it should come as no surprise. It is itself debatable. Why? Because it is extremely difficult to date ancient sites, certainly with any accuracy. It is entirely possible that our list of more than eighty sites will be revised in the future, either upwards or downwards. Such fluctuations are normal in the evolution of scientific knowledge. It is unlikely, though, that developments in our knowledge will disqualify very many of our eighty archaeological sites. So the corpus is necessarily much larger than that proposed in the original study.

We also emphasized in our article that this critical analysis of the chronology of the sites should apply to the evaluation of the age of the cannibalistic moment in our small cave in the Ardèche. The exact age of layer XV at Moula is in fact highly debatable. Only very few physical-chemical measurements have been made at this level, and the small number of analyses has in its detail thrown up fairly contradictory results, with ages varying between twenty and thirty millennia, and some substantial statistical uncertainties. A measurement with a margin of error of plus or minus ten millennia thus indicates an uncertain period falling somewhere between 90,000 and 120,000 years ago. Should we then consider the higher statistical limit or the median figure of these measurements? Whichever option we choose, almost none of the analyses indicate directly the main temperate phase, which we generally situate between 123,000 and 116,000 years ago. The researchers based their estimates on climatic indicators deducible from the analysis of small fauna – snakes, amphibians, rodents, etc. – and offer as evidence a mixture of fauna adapted to warm climates alongside cold-climate species. This mixture suggests the very start of the temperate period, a time when a number of species who lived here during the preceding ice age would still have survived.

This argument is not valid, however, as such mixtures of warm- and cold-climate species can be found at every archaeological level of the 6-metre-deep cave, over a time span of nearly eighty millennia. The site overlooks the vast valley of the Rhône, which provided the main migratory corridor connecting the Mediterranean to more northerly areas. The presence of warm and cold animals then fluctuated regularly and rapidly, as some species of a more temperate nature were able to head north along this natural migratory axis, while others, better adapted to cold continental climates, just as regularly descended south via the wide corridor. Reference to cold and temperate fauna is only a very poor chronological indicator and is not in itself adequate to locate our cannibalistic episode at the start of this huge climate change. The discovery of temperate and polar fauna in the same archaeological earth is a signature of the Rhodanian biotope, marked by continental conditions but still under a Mediterranean influence. So the argument looks particularly unsafe and should probably be abandoned, which leaves us even more uncertain as to the real age of the cannibals.

Are the levels in our cave really 120 millennia old? Or 100? Or eighty? The question of the positioning in time of the deposits of Palaeolithic prehistory is a crucial one and requires very great caution when you reach periods too old to be effectively measured by carbon-14 analyses. In theory, such methods do allow dating of bones older than 55,000 years, but in reality very few laboratories are able to reliably process samples older than forty millennia. Applying an exact date of twenty millennia to a site requires the implementation of very large corpora of dating cross-referring results from different methods which are generally much less precise than those of carbon 14. By using multiple measurements acquired using different methods we can construct robust statistical models with which we can then accurately position an archaeological sample in time. Very few sites can meet such rigorous criteria, which opens up the possibility of claiming, somewhat subjectively, that only a handful of sites in Europe date from this peak temperate episode. It is more likely, in reality, that there are numerous examples of such sites across the continent, but identifying them requires vast scientific programmes drawing on the skills of the best international teams. So on the whole these peoples of the forest are constantly eluding our grasp. Nevertheless, there are now a few large archaeological sequences that help us to address the reality of Neanderthal societies and understand how these populations adapted to radical climate change.

The temperate optimum lasted between ten and fifteen millennia. With temperatures up to 3 °C higher than they are today, it was the warmest climate episode on earth in the last 400 millennia. We are talking here about global temperatures and ocean temperatures. Locally, at our latitudes, the temperature difference could have been more marked, with seasonal values up to 10 to 15 °C higher than today. In the middle continental latitudes Neanderthal populations probably developed strategies adapted to the progressive spread of huge primary forests.

But if cannibalism was a direct consequence of radical climate change, what should we think about these populations that were incapable of adapting to the forests that covered Europe for more than ten millennia? Weren't the forests themselves reservoirs of huge biodiversity beyond compare with those of the preceding glacial periods in our latitudes? Were the Neanderthals unaware of these new animal resources or were they simply unable to change their ancient hunting traditions?

The variety of predators we have identified during this period – lions, hyenas, panthers, wolves, bears, gluttons – shows the existence of extremely rich biotopes. These predators need large resources of protein and could only have developed in natural habitats that were favourable to them.

Thus the whole scenario needed to be revised: whatever the age of the site, and whether the cannibals lived at the beginning or end of the temperate period, if the interglacial forests were so rich, if these environments were among the most abundant in game that Europe had ever known, how could it be that the Neanderthal populations, having managed to colonize virtually every biotope in Eurasia, had only been able to ensure the survival of their group by eating the flesh of their own dead?

Now, any society is prone to some catastrophic event that might lead the group to eat their own dead as a last resort, but this study was not basing its claim on some sinister one-off event from 100,000 years ago in one small cave in the Ardèche. Its argument was based on a global collapse of human populations across Eurasia, a collapse resulting from the inability of Neanderthal societies to adapt to rich forest environments.

This strange theory about cannibals in the forest simply did not stack up.

Ancestral Knowhow and Strategies

Episodes of famine are regularly documented in the history of human societies, and the signs of famine are visible in the growth lines of the Neanderthal teeth we discovered. But such indicators are not atypical and are commonly identified in analyses of teeth from hunter-gatherer societies, whether pre-historic or modern-day, such as the Inuit, for example. The Inuit today are the most boreal people on the planet. They have developed technologies enabling them to respond efficiently to extreme environments and ensure for their groups continuous access to food resources that are exploited by no other population on earth. In the far north, strategic errors are not an option; they potentially put the whole community in great danger. And yet the Inuit have never been known to practise any form of cannibalism, not even on rare occasions.

Traditional hunter-gatherer groups are generally found in large territories where they control and manage natural resources that vary greatly from season to season. The organization and activities of the group are carefully planned over several seasons, sometimes several years in advance. These traditional forms of organization, these rhythms of nomadism, draw on ancestral wisdom based on precise knowledge of the behaviour of animals, their migratory patterns, their habits and movements. Such knowledge is centuries, if not millennia, old and transmitted via the oral tradition, enriched from generation to generation through each group's intimate knowledge of its territory.

The organization of the group, its survival, its equilibrium rely on ancestral knowledge that is empirical, analytic and indeed scientific in its all-encompassing experience of the world. Certain Inuit groups divide the year into six, sometimes eight, seasons – start of autumn, autumn, start of winter, winter, end of winter, start of spring, spring, summer – which enables them to manage their natural environment in fine detail. The end of winter can be a critical season. An over-extended cold spell and the late arrival of game at the start of spring can have devastating consequences. Such phases, when the community might be placed in peril by a simple variation in climatic conditions, are well documented. Even though they involve exceptional circumstances, they recur from time to time, and the group's ancestral memory retains the experiences of their forebears. They are anchored in memories and give rise to strategies that an outside observer might regard as counterproductive, or even irrational. The vernal equinox and the end of winter are marked by large festive gatherings and sporting competitions: sleigh races, ball games, tug-of-wars. These late-winter gatherings are not just seasonal markers but provide an occasion for groups which have been isolated during winter to get together again and take advantage of the opportunity to dispense with winter surplus stock in great feasts, a communal consumption of unneeded provisions. If these great celebrations do mark the return of game and the days of abundance, they are not solely a communal expression of joy over the return of the fine days but represent one of the strategies for restoring the cohesion of the groups and to offload winter food reserves that are not needed except when an exceptional climate event occurs. The crisis, the unexpected, unlikely event, is anticipated and dealt with by strategies for managing non-perishable resources (dried, frozen, fermented), which otherwise will generally not be needed for facing their overly long winters.

If spring does not arrive, if winter extends even beyond the consumption of winter overstocks, the Inuit will move closer to neighbouring communities, whose winter shelter is well known to them. So it is the social link that allows communities to survive. Springtime celebrations enact the management of resources of protein and the social cohesion of the groups, dramatizing the two main pillars guaranteeing the survival of the peoples of the far north.

Although these populations base their logistics on an intimate knowledge of their environments, it is networks of friendship and alliances which ensure the longevity of the group. They represent the solid base that assures the survival of individuals and populations. It is these very networks of mutual aid which enabled the colonization of all the environments on earth. The resilience of human populations rests on a vast invisible network which to a large extent frees societies from the limits and constraints of the milieux they have colonized. Extreme famines, ones that bring about consumption of the dead as a last resort for the survival of the group, are almost completely unknown in hunter-gatherer populations.

Run! Run! These Things aren't Human!

Still, subsistence cannibalism does exist. It has been widely documented in our own history. But what are the exact circumstances that lead to such events?

By analysing situations of extreme human despair we can see what most of these anthropophagic crises have in common. They almost exclusively involve groups in transit, people who find themselves in unknown territory, in which the cannibals-to-be are unfamiliar with the resources available and have no network of mutual aid with the aboriginal populations.

In 1845, Sir John Franklin, captain of HMS *Erebus* and HMS *Terror*, led a polar reconnaissance expedition to locate the 'Northwest Passage'. The expedition aimed to open up a sea passage allowing navigation around the northern rim of the

Americas, in between the islands and archipelagos separating the Canadian far north from Greenland. That year, the ice seemed fairly thin, and the British were well used to conducting polar explorations. Franklin was considered to be an expert on the northern oceans. They took enormous supplies of food, enough to feed the 129 crew members of these two great ships for three to four years. Franklin foresaw more than one winter when the ships would be ice-bound during the worst of the season. But the ice did not release them from its grip even in the following summers, and in spring 1848 the crew finally decided to abandon their ships and their supplies of food. It was late April, summer was approaching, so the crew decided to walk to the nearest trading post on the Hudson Bay, 1,600 kilometres away. They would follow the Black River, which they thought would provide them with enough fish to cater for their needs.

Not a single one of them escaped alive.

In 1854, a Canadian cartographer brought back reports from the Inuit about cannibalistic activity in the far north. Many Inuit groups had been profoundly affected, giving eye-witness accounts of white cannibal demons with an inhuman appearance.

About ten or so men had reached the south-west edge of King William Island, where there was an Inuit encampment. The encounter was so shocking that the memory was still vivid some 150 years later. Some of the Inuit testimonies were collected in 1999 by Dorothy Eber. It was a woman who first spotted the terrifying, staggering figures with their blank eyes and blue flesh, incapable of speech. She ran as fast as she could back to the camp: 'Run! Run! These things aren't human.'

Over the last 170 years scientists have made regular discoveries of scattered vestiges of the ships, and remains of the crew are still being found to this day by polar expeditions. Analysis of the bones shows traces of flesh stripping, of their being broken to extract the marrow, and of the polishing that characteristically occurs when bones are stirred in pots to make stock. Every last usable calorie had been recuperated. As it turned out, the Black River was virtually devoid of fish; the Inuit themselves knew this region well and scrupulously avoided it, as there was so little food to be had.

We have here an example of cannibalism on a wide scale, involving the entire crew of two large British ships, documented by Inuit oral accounts as well as by analyses of bodily remains. Here is a macabre demonstration of how subsistence cannibalism can affect a whole large human group, whereby more than 100 people at death's door resorted to devouring each other. Such acts of despair are perpetrated by people who are trapped in unknown territories, stuck in extreme environments where the resources are insufficient for the needs of the group or are simply unknown. But these subsistence cannibal events are also above all related to isolation from one's own social group. There was no one there who could have come to the aid of these crews, whose cultural and moral values had been annihilated by the harsh reality of their situation. Even though the crew members from the *Erebus* and the *Terror* finally managed to reach another human population, the Inuit they encountered were so terrified by their dehumanized, bestial appearance that they escaped by sea to avoid all contact with the cannibal monsters that they did not even recognize as human beings. What the Inuit saw was dangerous creatures escaped from the annals of their ancestral mythology, no longer identifiably human. Flight was the only option in the face of these emaciated beings.

This story highlights that to our societies cannibalism

represents an unimaginable aspect of human behaviour, especially in the context of adventure, discovery and progress. It is rather as if the first missions to Mars ended up with the spaceships crashing and the crew eating each other. To the western mind there is an instinctive mismatch between the nobility of exploration and these stark extremes of the human condition.

A Neanderthal Ritual?

To my knowledge, no such acts of cannibalism to ensure the survival of an individual or a group have ever been recorded among hunter-gatherer populations occupying their traditional territories and living in environments that they have always known.

In our latitudes, climate change, no matter how abrupt, could not have fundamentally modified the structure of biotopes over the span of a human life. No horse-hunting Neanderthal of the steppes ever found himself suddenly transposed into a vast temperate forest teeming with deer. The Neanderthals of Moula dropped flints on the ground, mixed in with bone remains. These flints showed the exploitation of rocks over a large area of several dozen kilometres on both banks of the river. Such precise awareness of flint resources leads us to conclude that the group was certainly in its traditional territory and had marshalled all of its resources for a significant period of time. This was not an isolated group on the move through unknown territory, either unaware or unappreciative of its natural resources. So what actually took place in this little cave in the Ardèche? Were the cannibals of Moula really cannibals at all? Did they really eat the flesh of the bodies they cut up?

The flesh stripping is now more or less beyond dispute. All organic body parts were removed and separated from the bones, not just the muscles but the skin, the scalp, the brain, the marrow, the tongue. But was a single gram of this flesh actually eaten? And if so, how widespread was the practice in this population?

As we have seen, there is a great wealth of traditional practices concerning the handling of bodies. If endocannibalism and exocannibalism tend to overlap, it is because eating a human body is never an anodyne act, because it is always invested with great symbolic power, an ultimate dance of the living at the gates of death. In one sense a Neanderthal version of the medieval *danses macabres*. We are now entering the sphere of rituals and irrational behaviours common to all human societies.

And yet, the human bones themselves seemed to tell us a different story. These Neanderthal remains displayed abundant traces very clearly related to flesh stripping. Abundant, or rather super-abundant. As we look closer, we notice that half these human bones present traces of flesh stripping, a sharp discrepancy with visible traces on the animal bones with which they were mixed up in the same earth. The animal bones showed markedly fewer traces of exploitation: only one bone in four showed signs of having been cut with a flint. Even more surprisingly, the traces on the human bones were also visible on the anatomical parts with very little by way of nutritional value. Metapodes, phalanxes, collar bones and mandibles had marks which it would be very difficult to interpret in terms of exploitation of proteins. And, unlike the animals, none of the human bones had been burned. A comparative study of these remains, even a generic one, showed that humans and deer had not been treated in the same way. As if these acts of butchering pointed to two distinct events. The butchering of the deer and ibex showed the way in which Neanderthals exploited animal nutritional resources. But the testimony of the six human corpses seems to tell a different story. The more we explore the exact context of these historic events, the less *raw* evidence of subsistence cannibalism we find.

The Krapina cave in Croatia, unfortunately excavated too early, more than 100 years ago, threw up a very large number of Neanderthal human remains. It was at this site that the now much-discussed hypothesis of Neanderthal cannibalism was first mooted. It is true that the archaeological data are difficult to use, since the collections are for a large part mixed up, as is the case for the huge hole of the Moula shelter excavated by part-time Pierre. At Krapina, whole caseloads of Neanderthal remains were discovered. But without any precise context. They are composed of the vestiges of several dozen Neanderthals. Some claim that there are twenty-seven different bodies, others say eighty. In any case, this is the largest single Neanderthal collection, which could be of the same age as the butchered remains of our site in the Ardèche. One of these vestiges, a fragment of a Neanderthal face, shows amazing traces made by a flint. Around twenty parallel marks, difficult to explain in terms of the recuperation of flesh. Were the Neanderthals of Krapina and Moula expressing a very singular relationship with the vestiges of their dead? Have we here, in fact, hit upon a Neanderthal ritual?

When it comes to the Neanderthals, any evidence of rituals, symbols or actions of a spiritual nature is singularly lacking. Did they even bury their dead? That question may never be answered. Given the significant rarity of Neanderthal remains, the very question of the existence of burials divides the scientific community into two passionately, but robustly, opposed camps.

Did they ritually strip their dead of their flesh? Once again, the Neanderthals slip through our fingers. There is no simple answer. No obvious answer. To the best of our knowledge . . . Everyone has an opinion. Everyone draws their own conclusions.

Still, in the wider context of international research, our knowledge of these populations is evolving very quickly. In April 2021, Spanish researchers identified Neanderthal nuclear DNA preserved not in human fossilized bones but in the very earth of the caves. Genetic analysis of earth from the Galería de las Estatuas near Burgos, in the north of Spain, revealed an astonishing contraction of Neanderthal populations directly after the temperate period of the Eemian interglacial. The analyses showed a clear diversity of Neanderthal populations during the temperate optimum, 130,000 years ago. However, a few millennia later, around the hundredth millennium, the earth in the cave records only a single Neanderthal population. As if the return of the cold climate coincided with a collapse of human societies, and thereafter there was only one population occupying territories that formerly, in times when the climate was more clement, contained a much greater variety of populations. The picture that emerges from these genetic data is no longer one of a collapse of the human population at a time of climate rewarming, but the exact opposite. Just as with animals, whose biodiversity increased with rising global temperatures, in our latitudes the temperate climate was also favourable to the diffusion, diversity and expansion of humans, and it was in fact the return to glacial climatic conditions that brought about a collapse in biodiversity. In all biodiversities, animal and human, since in those far-off times there was more than one species of human too.

The conclusions we can draw from this are the opposite of the catastrophist theories relating to the impact of climate warming on Neanderthal populations. The idea of human bio-diversity obviously entails notions of cultural diversity and distant social diversities of which we still know next to nothing. We have too few sites to tell us about these ancient periods. And they are too badly dated. They record only short episodes, always discontinuous in time. In one place, an archaeological window allows us to look at what happened thirty millennia ago. In this other cave, a few scraps of information tell us about societies that were around 100 millennia ago. And elsewhere, a few bits of data give us a view of the populations of the eighti-eth millennium. But virtually no site allows us to precisely document the organization of human societies and the evolution of environments in this critical phase during which glacial world climates moved into a temperate period before plunging once more into a new ice age. Our cannibals of the Ardèche are decontextualized and cannot be precisely connected to the evolution of environments and societies in this part of Eurasia. And without this context, a full understanding of these episodes of human flesh stripping, whether rituals or ultimate acts of despair, remains out of our reach.

Did these Neanderthals ever practise any form of cannibalism or are we dealing with a ritualized handling of dead bodies? And is there any way of documenting, or even envisaging, the slightest trace of ritual activity among the Neanderthals?

4.

Rituals and Symbols

Neanderthals Interrogated

The question of the Neanderthals' nature is really part of a more
general question concerning human nature itself. A question
that has been with us since the dawn of time even if we haven't
been able to define it precisely. Human nature has haunted every
human society. It is central to all the fields of western thought,
from philosophy to psychiatry. Plato defined man as a featherless
biped; Diogenes the Cynic's riposte was to produce a plucked
chicken to illustrate the absurdity of this claim: 'Here is Plato's
man.' But this did not deter Plato, who then defined his man as
a biped without feathers or claws. You could go on for ever
removing or adding attributes to this poor chicken without ever
coming up with a clear definition of a human being, and perhaps in the end you would have to conclude that humans are
simply humans. To which some detractors might add that
humans are self-domesticated primates, basing human nature
on the processes that made humanization possible.

Death of the Great Ape, and a Final Farewell
to Our Prejudices

In all likelihood the question of how we see the Neanderthal
is only the gist, or a bad caricature, of these perennially

unresolved thoughts. Are they unresolved simply because, outside of our fragile mental constructions, there might be no such thing as human exceptionalism? Maybe humans never distinguished themselves from the animal kingdom. So are humans merely part of the great diversity of living creatures? The more our knowledge of animal ethology progresses, the more clearly it appears that neither tools, nor thought, nor laughter, nor empathy, nor love, nor social structures fundamentally distinguish our species from the multiplicity of other living beings. In all these areas, once thought to be unique to humans, research into animal behaviour now reveals profound connections between our species and other creatures. Far from radically setting us apart from the animal kingdom, our current understanding rather suggests we are one species among many. There is no longer a neat distinction, a clear, robust dividing line, but just degrees of reality in the definition of our human properties.

This idea is only shocking if we choose to remain trapped in our narrow views of what constitutes life. The philosopher and anthropologist Lucien Scubla throws a different light on the notion of humanity when he underlines that aesthetic display (such as plumage and song in birds, costume and musical rhythms in humans) can serve to identify species as well as human groups, emphasizing both the unity of the living world and the cultural diversity that we encounter among humans. It's a remarkable thought because it's so surprising. It adds the cultural expression of human societies to the melting pot of animal expression. And we are reminded that all human societies, in many different ways, mimic behaviours that we see happening in the animal kingdom.

In this context, the question of whether or not the Neanderthals were human starts to look different. As we have seen, there are two main schools of 'scientific thought' that hold opposing views on whether the Neanderthals were a fundamentally different humanity or whether we can project on to them all of the characteristics that are supposedly constitutive of our species. I put 'scientific thought' in inverted commas firstly because it is not the thought itself that is scientific but the tools and procedures used to analyse the world. And even if the tools themselves are scientific, they are only used in the second instance to develop and legitimize a thought. Thought itself is never scientific. Thought is free and, quite often, a prisoner of itself.

How does Lucien Scubla's theory impact on our understanding of the Neanderthals? Prehistorians try to distinguish or integrate Neanderthal behaviour in relation to our own humanity on the basis of a small number of indicators considered to be diagnostic of human nature as we understand it. Where once the dividing line between humans and animals was set at the level of chicken feathers, nowadays, as far as archaeology is concerned, nature and culture are distinguished on the basis of the emergence of symbolic thought.

But what is symbolic thought?

The core of this notion can be summed up fairly simply. A hat is an object; headwear is a function – the function of the hat. Headwear is more than just protection against the sun, it also sends a variety of conscious and unconscious messages giving information about the values of its owner and also expressing his or her status among people of the same group. This is not the ethnic group, or the tribe, or the nation. I am talking here about the larger group, that is, all those who instinctively understand the function of this headwear, all those who require no explanation to be able to grasp at a glance the difference between the crown of the king of England and

James's baseball cap. This instinctive and immediate understanding of the different meanings expressed here shows that simply by seeing these objects an individual is propelled into a thought space whose tacit rules are sufficiently powerful that they don't need to be verbalized. This function of the sign has been recognized and analysed by sociologists and philosophers for a long time. Such a type of symbolic thought is what is seen as a principal marker distinguishing the human from the infra-human. I use the phrase infra-human because, consciously or not, most studies place humans at the peak of evolutionary processes, thereby implying that all the non-human living beings are inferior from an evolutionary point of view. On the question of whether the Neanderthal was an infra-human, the archaeologist tends to produce a whole recipe list of criteria for determining whether symbolic thought did or did not exist in these populations. For us archaeologists, who have only fossil objects to help us understand our distant ancestors, such indicators might be art, adornment, graves, rituals.

However, Scubla offers a different viewpoint, one which reintegrates humans into the wider world of living beings. It is a broad, generous vision, but one which locates these principal indicators of human symbolic thought – songs, rituals, dances, adornment, ceremonies – in the wider animal kingdom. The question of Neanderthal burials, which has been an important factor in determining whether Neanderthals are older versions of ourselves, is directly impacted by this alternative reading.

Let us consider another example. Whether Neanderthals dug graves is a highly contested area of scientific study, because if there are graves, many scientists would have to accept the Neanderthals as version of us. Indeed, but why?

Because inhumation reveals an awareness on the part of the group of the unique character of each individual. An awareness of the irreparable nature of loss for the living. Graves therefore are one of the indicators that help us identify structural elements in the relationships between individuals in these human populations. We can then determine whether Neanderthals considered each individual as a unique and irreplaceable entity, which modifies our own view of the empathy, respect and sensitivity these humans felt towards each other. Self-awareness and the awareness of others. The grave marks my desire to protect those close to me, my loved ones. Whatever they bring me or don't bring me, therefore whatever it takes. It implies an 'I am' and 'you are' in these populations. There are other archaeological indicators which point to the existence of self-awareness, of awareness of others, such as the discovered remains of old, disabled or toothless persons who could not have survived without being cared for by the group.

Although the archaeological reality of Neanderthal graves remains a hotly discussed issue in the international scientific community, I personally have no doubt that the Neanderthals did in fact bury their dead and attempted to preserve their remains, following the many traditions developed by these societies over the millennia. I also have no doubt that these populations took the greatest care of their most vulnerable individuals – the young, the old, the invalids. But we must go further and push our reasoning to its logical conclusion. Do this relationship to the vulnerable and this relationship to the dead really document the deep mental structures of these people? Do such attitudes really represent something unique and specific to human beings? Something that would justify the claim that the Neanderthals are consistent with our own ideas and ways of being in the world?

Ethologists have shown that the empathy and sorrow directed towards the dead are shared and expressed by many animal species, from great apes to elephants via the dogs who sleep on the graves of their deceased masters. In 2010, an article by James R. Anderson in *Current Biology* described the death of Pansy, a female chimpanzee over fifty years of age who lived in a wildlife park. As her death approached, Pansy became short of breath. In her final ten minutes, the other chimpanzees drew close, took care of Pansy and caressed her as many as eleven times – uncommon behaviour in this group of hominids. When she died, the chimpanzees looked for signs of life, closely inspecting her mouth, lifting her limbs. After these inspections, an adult male attacked Pansy's body, an action interpreted by the scientists as an attempt to revive her or to express anger and frustration towards the deceased. The chimpanzees then abandoned the body, but her twenty-year-old daughter Rosy stayed by her mother's side the whole night, every now and again picking nits from her body. Rosy had never spent the night in this place before. The next day, the body was taken away by park staff, but for several days the chimpanzees refused to approach the spot where Pansy had died. All of these unusual events were filmed and documented in detail. The responses of this group of hominids included the particular care shown to the female before she died, the detailed inspection of the body looking for signs of life, the attempts to revive her or the flashes of anger on the part of the male, the grooming of her body, the vigil observed by the dead chimp's daughter and later the avoidance of the spot where the death had occurred. The study showed, unexpectedly, that hominids possess an awareness of life and death. Full consciousness of self and others, but also empathy and filial love.

Although we think of such actions and sentiments as fully human, we have to accept that they do not distinguish us at all from the animal kingdom but necessarily connect us with the distant origins of chimpanzees and humans – in other words, at least 13 million years ago, in a period when our populations had not yet separated, which attributes what we thought was one of the most human traits to a very distant ape ancestor from whom both humans and chimpanzees descended. This great ancestor, neither human nor ape, but in potential terms at least both, already had awareness of self, awareness of others, awareness of life and death, awareness of filial love, empathy. All of this involves associations of feelings and actions that are not specific to humans. To a lesser degree, different forms of altruism and empathy are widely documented in most mammals. The rat and the wolf, among many other species, express altruism and empathy, displaying equal awareness of self and other. The origins of this way of being are not particular to humans but can be found in this common ancestor of our species. If we consider the character traits uniting humans and wolves in their understanding of the world, the awareness of others must already have been present in our common ancestors more than 100 million years ago.

So it is rather dumb of us amazing modern *Sapiens* to be so astonished, or awestruck, at evidence that Neanderthals might have taken care of their own, whether living or dead. To be clear, I am not claiming that there is no distinction at all between humans and animals. Nor am I making out that there is no difference between us and the Neanderthals. What I am saying is that the questions concerning the nature of the Neanderthals, based on evidence of the existence of graves or of the care given to weaker members of the group, are fundamentally wrong-headed and will not in the slightest enable us to understand the structure of these ancient humanities.

The care that the Neanderthals bestowed on their dead arises from an ethology shared by all hominins and does not in any way link the Neanderthals to the conscious or unconscious conceptions that characterize our own humanity. And this evidence tells us that the archaeological facts are not under-exploited or over-exploited but more often than not misunderstood. An inaccessible sphere of prehistoric thought.

In archaeology, the facts presented are often more interesting than the way they are interpreted. But if the interpretative apparatus is weaker than the empirical reality of a past ethnography of these societies, what are we left with? If we don't see these sociological realities in black or white terms – as strictly symbolic or strictly utilitarian – but rather as a multitude of subtle realities in which human activities are not simply caricatured in binary categories, then this argument about the unlikely origin of human symbolism becomes purely superficial. The debate suggests that symbolism, a fragile notion, as we have demonstrated, is nothing more than an archaeological indicator of the existence in our distant past of actions that did not perform a purely utilitarian purpose. But doesn't all material production already inherently pass this threshold? It is an inescapable conclusion, unless we consider technology, from flint carving to aerospace, as simply a matter of blind mimicry, a type of reflex action, the learning of which would involve merely the acquisition of an inherited functional knowledge, devoid of any intellectual content. Who could seriously believe that?

Maybe the answer as to the origin of symbolic thought was already latent in the question itself. But in the end, who were the Neanderthals and which interpretations will uncover some of their reality? There may be no neat way to summarize Neanderthal thought, but there are facts that allow us to examine our archaeological archives.

If awareness of self and others, empathy, a relationship with death, a care for the living do not fundamentally differentiate us from any number of our animal cousins, and do not tell us of our human ancestors but of very ancient animal behaviours over millions of years, how can we get close to an understanding of the world of these past humanities? No more idealizing of Neanderthal graves. Their actions might reasonably be seen as just one of the many ethological variants rooted in an animal nature far older than any human form. The relationship with death, grieving, an understanding of the uniqueness of each individual – these are not distinguishing features. We archaeologists can no longer posit the emergence of a full humanity on the basis of these burials. It is not the care taken over a dead body that is distinctly human, not the act of inhumation, but its ritualization. But rituals, as we have seen in discussing all the different forms of anthropophagy, leave traces that are essentially intangible, fragile, debatable. The debate surrounding the existence and the significance of cannibalism is a good illustration of the complexity involved in analysing acts perpetrated by Neanderthal societies.

The theory of subsistence cannibalism is based largely on the hypothesis that Neanderthal societies collapsed in the face of climatic and environmental changes, but a close analysis of archaeological archives reveals that the particularities of these environments, their exact resources and the organization of the human societies that were confronted by them are essentially a mystery to us. Still, by examining this moment of rapid climate change, we might learn from the archaeological data a little more about the existence of Neanderthal rituals.

A Temporal Fault

Rapid climatic changes happened all round the world, and are well documented thanks to extensive core drillings in the deep seabed, or in the eternal ice of Antarctica and Greenland. Pockets of air trapped in the ice fossilize the water as well as small bubbles preserving samples of past atmospheres. These components of past atmospheres, held in fossil ice, give us the means to reconstitute changes in the earth's climate with a great degree of detail. Still, the exact impact of these climate changes on the continental landmass remains highly uncertain. The reaction of the biosphere is much less understood. While it is very easy to draw precise graphs marking changes in global temperatures, it is difficult to determine exactly how the biotopes reacted on different continents and at each latitude.

We have to search on continental land for fossilized evidence of the changes in natural environments. But away from the ice of the Arctic and Antarctic, the indicators are much more sparse and throw up only small fragments from different times disconnected from the real, progressive evolution of environments over several dozen millennia. We do find fossil evidence in ancient peat bogs, which over the course of time have trapped millions of bits of pollen, offering us a preliminary image of the past biotopes. Analysing concretions in caves – stalactites, stalagmitic floors – also offers a way into these climate archives, but here again only in small time windows of a few centuries or millennia. This provides some indicators documenting the reaction of the natural world to global climate changes. But this continental evidence is only ever discontinuous and often offers us little more than a very partial, or very local, snapshot of changes in environments.

Ice and ocean archives tell us that the temperate phase was divided into five distinct climate cycles, three warm phases interspersed by two cooler periods. So it was over a period of about fifty millennia, between 130,000 and 80,000 years ago, that human societies would have faced these forested environments. The first ten millennia of this 'warm era' were much hotter than climates today, but for the archaeologist these ten millennia are difficult to distinguish from the two other warm interludes. Analysis of coals, bones and pollen allows us to map climate changes in fairly fine detail, but for these periods of more than 80,000 years ago dating methods involve a high degree of statistical error spanning several millennia. Allocating archaeological layers to one or other of these temperate interludes is more often than not an inexact process.

During this long interglacial era, Australia and the Americas appear not to have been colonized by human societies. It is only on the continents of Africa and Eurasia that we can document the evolution of such distant societies in the face of profound climatic upheaval. And to my knowledge, across the ancient world, there is no archaeological sequence which records the whole of these reversals of climate. We would need a complete record, in a single archaeological sequence, to analyse in detail the strategies developed by human societies faced with probably quite rapid changes in their environments. Was cannibalism one of the extreme survival strategies on the part of some of these peoples of the forest? Or were these acts a sign of something more profound, fragile archaeological evidence of ancient rituals that we have not been able to recognize?

To understand the cannibals of the Rhône valley, to pin down this question of survival versus ritual, we need to understand the framework, the precise structure of these peoples of the forest. But the Rosetta stone of this long period of the last interglacial still needs to be deciphered. Deciphered rather than discovered, for it is possible that in the region of Mediterranean France we already have such a Rosetta stone, allowing us to see in detail the reality of the immense primary forests, their singular resources and the traditions practised by the peoples of the forests.

Our Rosetta stone is to be found near the northern flanks of Mount Ventoux, the giant mountain of Provence. Ventoux is the highest Mediterranean peak on the edge of the Rhône rift. This monolith occupies dozens of kilometres of the Rhodanian landscape and has huge reserves of exploited flints of various ages, but also some Neanderthal sites of the utmost importance. These distant Neanderthal traces remained essentially unexplored until 2008. There was initially little evidence to go on, which is why the site did not really attract much attention from prehistorians, who were probably unwilling to undertake extensive research on the basis of a single flint found in the 1960s, at ground level, in a sort of horizontal fault at the foot of a soft limestone cliff. This fault would reveal a veritable Rosetta stone of the peoples of the forest. In less than a decade of research we would progressively uncover an impressive archaeological sequence which fossilized over a depth of 12 metres the principal climate phases of this temperate period.

The original flint found at ground level was kept in the natural history museum in Avignon, the Muséum Requien, in a small cardboard box with a handful of bones found in the same fault in the rock. A fault, or more precisely a sort of fissure, 50 centimetres high, with two vast smooth slabs as its ceiling and floor. A very small collection of bones, barely a dozen fragments, but representing an amazing diversity of fauna. Each piece of bone belonged to a different species, a veritable cavalcade of animals, all of them clearly forest fauna. There were wolf, two distinct species of bear, but also lynx, horse, deer, bison, ibex and turtle. One small bony fragment seemed to come from a cave hyena. That is quite a haul from just a few splinters of bone.

The combination of wolf, cave bear, brown bear, lynx, hyena and ibex is remarkable. This is the Vaucluse, and we know of only one other site where such a combination has been found. The presence of cave bears is almost unique in Provence. The single flint discovered provides very little in the way of information about the artisanal traditions of the person who discarded it, but analysis showed that it was almost certainly a Neanderthal. A Neanderthal tool, bones revealing a forested environment and a great biodiversity. It had to be worth exploring.

The site is in the magnificent setting of the gorge of the Ouvèze, a tributary of the Rhône on its left bank. The gorge is just a few kilometres long, but what a backdrop! The river flows gently between immense luminous yellow cliffs. To get to the location where the flint was discovered we passed under huge natural shelters, where the cliff forms impressive over-hangs largely overgrown by old ivy vines. The cliff walls are covered in thick mosses, and there are massive rock concre-tions overhead. Early research between the nineteenth and the middle of the twentieth centuries revealed that numerous pre-historic populations took shelter here: Mesolithic fishermen, a dozen millennia ago, Magdalenian deer hunters a few millen-nia before them, and there were a few indications of much more ancient populations, which is what drew us. Our flint, from the interglacial period, is ten times older than the other prehistoric occupations recorded in the gorge of the Ouvèze. The timescale is dizzying. But, as I advanced along the ridge, I could also understand why researchers are no longer interested in the prehistoric occupations that are fossilized at the foot of these cliffs. The ground is littered with clusters of large blocks, between 3 and 10 metres wide. Limestone is extremely soft; it is made up of sands from ancient fossiliferous seabeds. The rocks are studded with millions of fossil shells, sea urchins, corals and enough shark teeth to last you a lifetime. These fossiliferous rocks are fragile, they disintegrate, releasing the sands of which they are composed, a slow millennial trickle of fine sand, but studded by meteor showers of massive blocks several metres across. To search here you have to weave your way among this chaotic pile of large rocks. The surfaces cleared by the excavation are in reality very quickly filled by these mineral deposits.

The archaeologists' ambitions foundered on the harsh reality of these rocks. We arrived at the site. Little more than a crack, in fact, which is why it is called 'low cave'. Very low for a horizontal fissure about a dozen metres long and whose opening, between the ceiling slab and the rock base, is no more than 50 centimetres tall at its highest point. Its other name, the Great Flea Shelter, isn't exactly encouraging either, and we came back from our first expedition covered in insect bites. I insinuated myself into the fissure to try to get an idea of where the handful of bone fragments and the carved flint could have come from. I crawled as best I could, but there was not much to see. The floor was covered by a chaos of blocks between 50 cm and 1 metre across. No sediments to dig in, but a heavy dusting of rocks along the fifty or so metres of this fault. Crawling to the edges of the cavity, I discovered a small stalagmite. Within its concretion a fairly massive bone was visible, bison perhaps, and a large piece of coal, well preserved.

Excellent, the concretions must have fossilized the ancient soils, now no longer here.

On my return from this little expedition to the Ouzève I decided to undertake some research in the Great Shelter with a small team and a budget close to zilch. Think about it, who would want to put money on a crack in the rock and one concretion? But I had just been recruited by the National Centre for Scientific Research (CNRS), and my small salary as a young researcher would be enough to cover the essentials. So with my team of friends and relatives, I set off on the adventure. I love scientific projects that, as in ethnography, require nothing more than that you immerse yourself in your subject. You just need to get up one morning and go, and you are underway.

The science here is in the process: simple, logical methods on how to string together your thoughts, use your ten fingers, find a new way to look at the world. Make sure there's enough to eat, a place to sleep, and above all take with you bags and bags of passion and energy. The strategy? To evacuate the blocks and finely sift the few sandy sediments that had trickled between the stones. We gathered them up carefully and sifted them through a very fine sieve with quarter-millimetre holes. I wanted to be able to safeguard the tiniest traces of the presence of Neanderthals and of their environments, and the smallest millimetre of rodent tooth might deliver a modest yield of information. But it was a gigantic task. To obtain a few rare buckets of this first prehistoric sediment we had to lift quite a few blocks: to get ten or so buckets of sifted sand we had to remove 75 cubic metres of rocks. As we were in a narrow fault, the dry atmosphere quickly became clogged with dust. It made our eyes water, it got in our noses and throats, and after a few effortful days of crawling, pushing, breaking, our handkerchiefs were flecked with blood. Fine sand suspended in the air is essentially composed of the siliceous dust of fossilized shells which is particularly aggressive to the respiratory tract. The shark's teeth we encountered by the barrel load are even more abrasive. Despite our protective gloves, one of our team almost lost a finger when he tore a tendon on a shark tooth protruding from one of the blocks. The entrance was too narrow to remove the largest blocks. We had to break them up with a hammer and chisel in the narrow, confined space and then roll them up the rising cave floor before evacuating them down the slope below the site. Weeks of breaking up and removing blocks, bleeding from the nose and scratched all over.

After a month of this extreme regime, without a break, beneath the expanse of blocks we finally found not the rock floor but, to our surprise, a very yellow soil composed of hardened sand. The pile of limestone blocks had effectively preserved the archaeological layers by sealing them in. There were potentially some real archaeological layers still preserved, and not simply some vague concreted relics at the sides! Cue cries of joy in the cavity, where it was now almost possible to stand upright. Our spring campaign was nearing its end, but we had discovered something wonderful, which gave us great hope. We would be back in the autumn, without any additional funds but with the hope of discovering something.

In October the gorge was bathed in the light of an Indian summer. The site overlooked some old poplars whose leaves turned a dazzling golden yellow for a few weeks. It was as if the light came as much from the earth as from the sky, but maybe that was just how we saw it, excited by the thought of new discoveries after so much effort and the long weeks of waiting. We started a survey of 2 square metres of the lovely light sand. We soon began to find bones, incredibly fresh, as if the animals had only just decomposed in the cave. Bird, turtle, deer, beaver . . . so many beavers! More beavers than at any other prehistoric site in continental Europe. Complete mandibles, superb. The bones had distinct traces of cutting. Then lion and the remains of a lion cub. And lynx. Wolf, a whole row of vertebrae in quick succession. Incredibly, each vertebra of the wolf bore a series of parallel stria, very fine but distinct. They had been inflicted using a flint tool. This wolf had been chopped up to extract its sirloin. The lynx showed neat marks of butchering, quite visible at the end of its paws, on its metapodials, parts that aren't especially appropriate for getting meat.

And then the first flint turned up. It took our breath away – lying flat in the sand, almost like new. As if it had just been carved. Extraordinarily fresh. I have led missions from the Equator to the Arctic Circle, and even in the frozen lands of the far north I have never seen anything as freshly preserved as this. It was as if this flint had just been carved; its deep caramel colour showed no sign of alteration; its cutting edge, finer than a razor blade, seemed free of splinters to the naked eye. It was placed next to the bones, where it had been discarded more than 100,000 years ago – the date was not in doubt – along with these temperate animals. Later analyses would confirm this. The evidence of the coals indicated a very rich temperate forest; the thousands of remains of rodents, snakes, amphibians and a profusion of snail shells recovered from the sands allowed us to reconstitute a highly forested environment and an immense biodiversity. But how had these bones and flints been so well preserved, as if frozen in time, as if time had stood still in this crack in the rock? It was not so much a fault in the rock as a fault in time. Bones and stone tools were enveloped in this very yellow sand which had come from the decomposition of the rocks of the ceiling. Enveloped in the same sand that had petrified the vestiges of sharks and fish in the Miocene sea some 15 million years ago. The incredible preservation of flints, bones and coals was the signature feature of this fossiliferous sand. Its preservative properties had worked equally on the Miocene vestiges and the remains of Neanderthal hunts.

Our little survey turned up a small but remarkable collection of these Neanderthal vestiges. The longer our work continued, the more the properties of this time-stopping sand astonished us. We found deer antlers as beautiful as the day they fell, a complete turtle shell, the only known example in the whole European Pleistocene, a small fireplace created by the hunters, with ashes, coals, stones reddened by the fire and twigs of wood. Not so much coals as unburnt wood, small twigs simply mineralized by the sand. We had never witnessed such a process of preservation. It had been an enormous task; the milieu had been gentle with the prehistoric vestiges but harsh on the archaeologists, knees and hands scratched daily, breathing difficulties, huge amounts of rock to break up and roll up the slope for removal. We also had a responsibility to leave the surrounding natural site as we had found it and an agreement with its owners that we would not disfigure this side of the valley with piles of rocks. So we had to build dry-stone walls, beautiful lines of blocks, much as our ancestors had done from antiquity to the nineteenth century.

Despite the forbidding conditions we decided to continue our work and returned for two months every year for eight successive years. Year after year, as we dug up the soil, we gained a greater understanding of the properties of this fault in the rock. The large floor slab, covering the opening of the cavity over its 10-metre length, is not the base of a crevice but a section of ceiling that had collapsed at the entrance to the cavity. The archaeological levels seem to have been pressed down under this huge slab. We needed to explore the entrance area of the cavity, in which the Neanderthals could have been active and left important vestiges. However, this time we weren't talking about moving 50-centimetre blocks but a huge piece of rock measuring more than 4 cubic metres, so I turned to the caving teams for help. Some of them were experts in rescuing cavers trapped underground. They are masters in using explosives and have been known to set charges just centimetres from the face of a trapped caver. Their expertise would be put to good use. Underneath the huge entrance slab the archaeological layers were as well preserved as those in the main chamber, and after the removal of dozens of cubic metres of rock we were able to see the original morphology of the cavity we were working in. We realized that the cavity did not have solid walls at any of its edges. Rather than natural cave walls, we discovered that its sides consisted of a mixture of concretions and piles of blocks.

This was neither a cave nor a shelter, we were simply underneath a very large rock and at the entrance of a huge maze consisting of a chaos of collapsed megablocks. Some of the blocks were as much as 20 metres across. These people lived at the entrance to the gigantic natural chaos. In the gorge of the Ouvèze the high cliffs towering over the river crumbled away into a huge pile of rocks long before humans appeared on the scene. The gaps between the blocks created a remarkable underground labyrinth – a three-dimensional labyrinth. The skein had no limits other than the edges of this massive rockfall. The gaps attracted and accommodated humans and carnivores and gradually filled up with sediments, fossilizing and trapping the bones and flints left by their successive generations of occupants. So the site was by no means limited to the few dozen square metres that we had explored under the first block. The whole flank of the gorge contained fossilized traces of our distant peoples of the forest. A search of the edges of the first cavity allowed us to recover flints and bones. In whatever direction we looked, we could see archaeological material. The search area seemed gigantic, and we were going to explore it using our own skills combined with those of our caver teams, who were familiar with underground exploration.

We reconnoitred around the site of one of the many entrances created by the chaos of blocks. These entrances were full of rocks and sediments with fossilized traces of the Neanderthals' presence. Other possible points of access could be seen above and on the sides of our megablock. After digging only a few centimetres around these exterior lateral entrances we immediately turned up vestiges of temperate fauna. We were, however, 12 metres above the soil we had earlier explored in the main chamber. We began to understand that the archaeological recordings in these 12 metres exclusively related to the temperate interglacial period. For the next six years we would attempt to join up the upper and lower parts of the maze. We set up wagon and zipline systems; every piece of archaeological evidence was precisely located, each handful of sand was sifted down to a quarter of a millimetre. There was archaeological material everywhere, all of it marvellously preserved. We advanced progressively towards the jumble of the underground space, but our slow advance was hampered by piles of large blocks that needed to be broken up and hauled out on hands and knees, while we recorded all the wealth of archaeological, geological and topographical information that we turned up. It was hard work, but we were beginning to really understand the structure of this huge excavation.

After six years labouring away underground, joy unbounded:

one Sunday evening, with the help of our chief pyrotechnician, Frédéric Chauvin, we finally made a direct connection between the summit of the infilling and the back of the first chamber that we had explored in 2008. For the first time we could grasp the full extent of the site. Our work had finally succeeded in joining up the different zones of archaeological exploration. Now we could see the whole 12 metres of archaeological records. We could then discern that there were fifteen distinct archaeological layers. The vast collections of bones and coals we discovered pointed to exclusively temperate environments, from the base to the summit of the vast subterranean archaeological infill. Given the importance of the discoveries, we were awarded two prizes by *National Geographic*. This enabled us to install walkways and security deep down in the heart of this impressive underground maze.

From the Peoples of the Forest to the People of the Deer

We now broke down the archaeological sequence into broad phases defined by the most representative fauna that we found in each: beaver, turtle, deer, hyena. But we found no less than sixty-one different species, from the elephant to the lion via the grass snake: one of the best examples that we know today of biodiversity in the prehistoric world. Analysis of the coals pointed to a very densely forested world. It turned out that the site contained fossils only from the interglacial period around 100,000 years ago, with a markedly warmer phase at the base, which we interpreted as the first major climate shift, the temperate optimum, confirming the hypotheses established in 2010 in the *Journal of Archaeological Science*.

But it remained to be seen exactly where we were in time.

Our partners at the universities of Oxford and Adelaide set up a vast corpus of datings covering all the archaeological phases we had identified. The results were beyond our wildest dreams. The soils contained fossils of settlements dating from 123,000 years ago for the oldest to 80,000 years for the most recent, 12 metres above. This time, there was no longer any doubt that the deepest levels of the cavity recorded the earliest moments of the interglacial. A hundred and twenty-three thousand years ago, temperatures were at least 2 °C higher than temperatures today. The successive dates matched perfectly and irrefutably our succession of archaeological levels, the ages becoming more and more recent the higher we moved up through the archaeological layers. For the first time the whole of the interglacial period had been recorded and documented in a single location – the changes in both the forest biotopes themselves and the strategies of the Neanderthal populations who inhabited these environments. The sequence represents probably the most complete archaeological record of the various phases of the interglacial period.

What this taught us about these peoples of the forest is of fundamental importance. First of all, we noted the presence of carnivores in the forest environments. From cave hyena to wild cat, all known species of carnivores were represented. Represented in great numbers, in fact. The profusion of carnivores indicated that this biotope was able to support them. A wolf eats 5 kilos of meat a week, while a hyena needs 4 kilos a day and a lion 9 kilos, though it might eat as much as 25 kilos on days of feasting. They are all social animals, living in groups, rarely in isolation. Spotted hyenas today live in large packs of as many as eighty individuals. Analysis of pollen found in fossilized hyena droppings indicates a very rich forest landscape. Yet this species is generally considered to be adapted to wide-open grassy spaces. It seems that the hyenas here were especially well adapted to the great temperate forests.

Some of the subterranean levels of the cavity are layers up to 2 metres deep not of sediment but almost exclusively of fossilized hyena droppings. Very large populations of these carnivores moved into the gorge and made it their territory. In all the phases of this temperate period there seemed to be exceptional resources of animal proteins available in the immense forests, which sustained remarkable densities of carnivores. The great Eemian forest was also a provider of resources. The sixty-one species recorded pointed to an incredible biodiversity. The humans took full advantage of this biodiversity. They confronted and exploited all these species, including the large carnivores. The bones of wolves, bears and lynx bear the traces of butchery. The specific nature of the cuts suggests that sometimes it wasn't just meat they were after, but also skins. In common with the beaver, these large carnivores had skins of a very high quality. Analysis of the cutting edges of some of the flints confirms fur-making activity and the use of ochre to disinfect and soften leather. The fine blades of these flints still bear red stains, visible to the naked eye.

Although the subterranean maze was regularly occupied by carnivores, humans also used the cavity over the course of time. Even in its deepest part, in the hyena dens where the soils are littered with rhinoceros and bison bones and hyena droppings, to our surprise we found a single flint object. A solitary object in this subterranean space. A single flint point, finely carved, whose acuminated tip had splintered on contact with a bone. The Neanderthals had gone into the underground space to hunt cave hyenas. In the heart of their habitat, in their very dens. And while hunting wolves might seem less risky at first sight, it is no less surprising. How do you catch a wolf unawares?

It can smell you and hear you coming from hundreds of metres away, even before you have spotted it. It is highly intelligent and extremely vigilant – it simply cannot be taken by surprise. It can reach a speed of 60 kilometres an hour in a matter of seconds and would beat our Olympic champions hands down in either a sprint or a distance race. In many traditional societies wolves aren't hunted, they are trapped.

We saw evidence of the exploitation of an extreme variety of resources and the mastery of very distinct strategies, probably ranging from face-to-face combat with the fearsome cave hyena, armed with a strong lance, to the trapping of the most elusive animals. They ate the meat, they plundered the finest skins. But the biggest surprise of all was in Alpha unit, in the so-called deer formation. This is at an underground level where there is no sunlight. The only illumination comes from torches or fires. The walls are still marked by fine grey deposits; soot deposited there a little more than a hundred millennia ago is still perfectly preserved on the rocky edges of the cavity. At this time the underground maze was not occupied by carnivores, so all the bones in the 'deer formation' were brought here by Neanderthal hunters who carried back entire red and roe deer to this part of the cavity. A precise analysis of the bones in this unit was conducted by Cendrine Beraud of the University of Bordeaux. It threw an unexpected light on these populations.

Typically, analysis of the prey of Palaeolithic hunters reveals the exploitation of game of all ages and both sexes. These hunter-gatherer populations predominantly hunted adult animals, with a smattering of young and old deer, and males and females were hunted in equal measure. But the Alpha layer did not conform with this random distribution of sexes and ages of the hunted deer. Here, there were no young or old deer and no females. The hunters clearly focused their attention on the killing of fully mature male deer, to the exclusion of the other categories. There were a large number of hunting trophies of skulls with antlers still attached in these collections. This was a virile hunt. A hunt focused on the strongest individuals, the ones hardest to kill. The animals were probably killed using a lance. This was face-to-face hunting – and the flint weapons found in this archaeological layer were long tapered points, massive and richly worked.

Neanderthal Rites of Passage?

The profile of these hunts should make us stop and think. As well as undertaking dangerous hunts of bear and lynx and complex ones such as that of the wolf, these humans also ventured into the dens of hyenas to confront these large carnivores on their own terrain. In the deer formation, the systematic selection of mature males was not related to any food needs. Ethologically, in their natural habitat, males and females generally live in separate herds, but the male herd includes younger and older individuals, even if a few older stags live a more solitary life. We have here a singular exercise of choice that is not easy to reduce to a basic logistical or economic imperative on the part of these Neanderthal hunters. The exclusive search for adult males probably represents something else.

There is a very rich ethnographic corpus, gathered across all continents, that shows us that for humans hunting is never simply reducible to a simple search for proteins. It is always ritualized, invested with codes that go way beyond the rationale of feeding the people. The hunt is never just a matter of logistics, to do purely with food. Humans are above all irrational matter. Although humans and animals overlap at many points, this singular way in which humans kill their prey projects us far beyond the animal kingdom, into a cultural elsewhere that is profoundly codified. No large carnivore has ever limited its hunting to a systematic selection within a single species based on gender or age. This is a distinctly human characteristic. And especially surprising. The ethnologist Bertrand Hell, in his book *Dark Blood: Hunting, Forest and the Myth of the Savage in Europe*, has demonstrated that since Antiquity, and in different European societies, the hunting of deer, and especially large males, is never anodyne. It is generally a solitary activity, involving a direct confrontation between the hunter and the deer. These traditional one-on-one hunts are not interested in females or the young or old, but exclusively in the most powerful males. They are exclusively masculine hunts and mark in a very ritualized and codified way the entry of the hunter into manhood. Hell outlined the transformation of the hunter into a man-deer through direct confrontation with his prey, as if the hunter had himself become a deer confronting a sexual rival. And even if the deer was eaten, the formal reason behind these hunts was none other than the direct confrontation between the deer and the man-deer. This rite far surpassed banal, rational, statistical requirements, the acquisition of flesh for sustenance.

We saw in our discussion of burials that a consciousness of the other and the grief and care expressed at its death do not distinguish us from the animal world, but quite the opposite, they make us part of it. Now, for the first time, we have hit upon ritual, that is to say on characteristics and behaviour that are uniquely human. One could, of course, counter this view by looking at animal behaviours and trying to find ritualized behavioural patterns, and I have no doubt that certain ritual forms do exist in animals. Research on animal behaviour has long shown that several species possess actual traditions which they transmit across time, from generation to generation. Animals too go beyond brute animal behaviour, analyse the world, invent and develop strategies and transmit these strategies to their descendants. In other words, they develop strategies that are not unique to their species – magpie or wolf – but unique to a particular family of magpies or wolves in a particular territory. It is fascinating to realize that animals themselves transcend simple ethology and are inheritors, like us, of traditions, and literally bearers of transmitted cultures which are proper to them. Field rat, city rat: each has their own values.

Our distant Neanderthal, that extinct humanity, is just as much part of these structures as we are. It is up to us researchers to follow through on this and no longer limit ourselves to positivist, mechanist, statistical, quantified, rational approaches which negate human nature – a scientific deviance from our ways of seeing the world. This type of positivism, which analyses only the superficial mathematically measurable structures of humanity, is a failure, an obstacle to thought. It is a way of denying human nature and the animal logic rooted within humans. It hides behind graphs, measurements, tables in order to avoid looking human nature squarely in the face. It is rigorous, of course. But this rigour is that of a statistician counting the number of drops contained in the ocean. It is cautious too. But it is a timid caution.

By limiting humans to their quantifiable rationality, this positivism has scientific pretensions. More deeply, and more sadly, it displays a simple refusal of thought. It is a putsch of the hard sciences against the human sciences. It projects on to, and into, humans an inhuman nature. We must look at and analyse humans in a human way. A form of participative ethnology of the old extinct humanities: neither projection nor fantasy nor construction, but an acceptance of the reality and the richness of the living creature. And an acceptance of the logic expressed in past behaviours.

Neanderthal societies reacted according to their own codes in hunts that we have every reason to believe were already ritualized, even if they took place more than a hundred millennia ago. We have progressed from the peoples of the forest to the people of the deer. We still know very little about these peoples of the forest, or about the astonishing people of the deer, whom we encountered for the first time in our subterranean maze in the gorge of the Ouvèze. We still have a long way to go before we understand these societies. Does the archaeological evidence here indicate the existence of rites of passage into adulthood?

We have seen that the origin of symbolic thought, which we like to think is unique to humans, but probably distinguishes us only by a matter of degree, was dependent on a number of archaeological indicators resembling a bad list of ingredients. Take one pinch of burial, add three pinches of bodily adornment, a soupçon of movable art, a smidgen of rock art, leave to simmer for 100 millennia and you end up with a tasty modern human ready to express deep symbolic thought.

Unlike symbolic thought, the argument in favour of ritual does not depend on discovering exceptional objects – a grave, a piece of jewellery – but on an analysis of the general structures of human societies. Here we see, perhaps, rites of passage to adulthood; there, the ritualized forms of cannibalism. Now that we know our peoples of the forest a bit better, let us return to our cannibal cave. Whatever the actual epoch in which these events took place – whether it was 125 millennia ago, at the time of our beaver hunters of the Ouvèze, or 100 millennia ago, at the time of our people of the deer, is it at all believable that Neanderthal societies of the Rhône valley were unable to adapt to life in these immense forests? Is it believable that they were unable to find game to exploit when, just three days' walk from there, these same forests were sufficiently rich to sustain the greatest animal biodiversity ever documented in Mediterranean regions? Could these forests have really supported whole families of lions but not humans? When anthropophagous cravings are not the mark of extreme hunger, Lévi-Strauss noted that they could only be down to a magical, mystical or religious cause: a positive form of anthropophagy.

Instead of a search for significant objects or actions, a structural approach to these past societies gives us an insight into the possible existence of ritual. Rites of passage into adulthood, ritual handling of the bodies of the dead. This is what defines the creature as a humanity. But our humanity? I wouldn't bet on it.

5.

Neanderthal Aesthetics

Did the Neanderthals have an aesthetic sense? I am quite sure they did; it is expressed in numerous ways in their artisanal work. So many of their objects display a quest for balance and elegance well beyond mere functionality. These qualities of Neanderthal tools are usually given a merely cursory attention, as if their aesthetics were merely a by-product of their functionality. This has consequences, as it fails to tell us anything about the structure of the Neanderthal mind.

Neanderthal Art and Volatile Thoughts

And yet a large section of the scientific community will tell you that we have now discovered Neanderthal art. Movable art, body decoration and even rock art, the great art of the caves. In its mental structures the creature is basically the same as us, and only our prejudices prevented us from seeing its true humanity. A victim again of that caveman image.

Let's be clear, this 'Neanderthal art' is based on very little. Ambiguous traces, even more ambiguous than the rituals of life and death of the peoples of the forest. Raptor claws, large feathers of birds of prey, drilled shells and a Neanderthal necklace.

In 2014, together with a few Italian colleagues, I published a study on eagle claws found at two sites dated to around fifty millennia ago. The first of these claws came from the cave of Rio Secco in the north of Italy; I discovered the second in the Rhône valley, in the Mandrin cave, a remarkable archaeological sequence with fossils from eighty millennia of human settlements, from the peoples of the forest to the Neanderthal extinction.

Three years earlier, the same Italian team had published a study that caused quite a stir. Analysis of bird bones found in the Fumane cave on the southern flanks of the Venetian Pre-Alps showed that these people gathered remiges, or flight feathers from the wingtips. Research into these large feathers was based on analysis of the bones of several species of birds: bearded vulture, black vulture, golden eagle, red-footed falcon, wood pigeon, red-billed chough. These were discovered in soils dated to 44,000 years alongside hundreds of flint tools that were unquestionably Neanderthal. At this point in time, *Homo sapiens* had not yet reached western Europe, and there is every reason to think that the collection of feathers had nothing to do with functional or food purposes. The Neanderthals supposedly collected them simply for their beauty. Probably as bodily adornment. The media seized on the announcement, and soon social networks were awash with images of Neanderthals sporting large coloured feathers. Our image of the creature was turned upside down again. Not so much Neanderthal as the Last of the Mohicans.

These findings seemed to connect up with the sensational announcement in 2010 of pierced shells having been discovered in the Neanderthal excavations at Cueva Anton and Las Aviones in the south-east of Spain, shells which were themselves interpreted as Neanderthal jewellery.

Four years later, the discovery of our raptor claws seemed to confirm the dandyish image of the Neanderthal, adorned with those avian accessories that many traditional societies invest with symbolism. Another five claws had already been discovered in other French Neanderthal sites. To date, around a dozen of these claws have been located, including the discovery in 2015 of another eight raptor talons in the site at Krapina in Croatia. Although there is no reliable dating for the talons excavated at Krapina back in 1905, there is little doubt that they are Neanderthal.

These findings from Neanderthal digs around the Mediterranean rim were immediately interpreted as proof that Neanderthal societies expressed themselves visually and symbolically, with evidence from the 120th millennium through to the extinction of the Neanderthals a little over 40,000 years ago.

Illusions Turn to Dust

In February 2018, a cover story in *Science* reported the most spectacular of this series of discoveries, one that would finally put paid to outdated images of the Neanderthals and restore them to their full humanity. Our humanity.

What the study announced was nothing less than the discovery of the most ancient rock art in the world in three Spanish caves: Ardales, La Pasiega and Maltravieso. Various signs painted in red ochre and even a negative hand, which showed that cave art was practised earlier than 67,000 years ago. It was an amazing discovery; the sixty-seven millennia here were established by the age of the concretions covering the ochre drawings. The dating is in fact a minimum age; the Spanish cave wall drawings could be even more ancient. Even at 67,000 years, this is twice as old as the Chauvet cave, the most ancient decorated cave in the world, and these three caves take us back to a time when the Neanderthals were the only inhabitants of the Iberian peninsula.

And we continue to find Neanderthal art even further back in time. The first Neanderthal accessories reported, in 2010, were relatively recent – between the forty-fourth and the fiftieth millennia – but more recently discovered vestiges of this type have been much older. The bar has gradually been pushed back from the fiftieth to the seventieth millennium, then to the 120th. It felt as if Neanderthal art had existed for ever. As if it were integral to this humanity, for as long as Neanderthals had been Neanderthals. But then, in 2019, a scientific team in the Qesem cave in Israel discovered feathers of swans, pigeons, crows and starlings in archaeological levels dating back more than 420,000 years. Well before the first Neanderthal population and very long before any *sapiens* population.

Does that mean that such distant past populations, who existed well before the Neanderthals, now seem more like another version of ourselves? There is something not quite right here.

How was it that the walls of these caves were covered in powerful symbolic representations when Neanderthal artefacts unearthed over the last 150 years have shown no sign of this Neanderthal sensibility before? But what about the Neanderthal's necklace? The raptor claws, the pierced shells, the large coloured wing feathers?

Closer analysis turns these illusions to dust. Not a single one of these objects is at all manufactured, none bears the slightest trace of a deliberate artisanal modification, and, after a century and a half of searching, we haven't in fact found the first hole of the first Neanderthal ornament. We are not dealing with concrete, tangible, objective facts here, but rather interpretations, projections and constructions.

Birds such as magpies that collect shiny or brightly coloured objects perform just as well at this particular game as Neanderthals. Bower birds of New Guinea and Australia actually collect hundreds of objects – stones, flowers, shells, feathers – of different shapes and colours in order to assemble them into aesthetically pleasing, eye-catching arrangements around the entrances to their nests to attract females. That's more than the Neanderthals ever did; our creature is outdone by the very birds it supposedly adorned itself with. Although it has long been known that Neanderthals gathered unusual objects – crystals, fossils, coloured rocks – similar behaviour is already evident three million years ago in the form of a small, unusually shaped pebble of jasper collected by Australopithecus in the Makapansgat site in South Africa. In ancient ethological terms we are pitched somewhere between birds and Australopithecus; in no way is this uniquely characteristic of humanity.

Should we then chuck this Neanderthal necklace in the bin as just another projection of our fantasies about the creature? Really?

But what about the pierced shells? Under a microscope these shells revealed polishing on the surface that might result from hanging them on a thread or on vegetable fibres; the study in question suggested that the shells might have rubbed together. Surely that beats bird and Australopithecus hands down?

Not so fast. These holes were in fact made by . . . crabs. All the Neanderthals did was collect handfuls of shells from the beach, some of them naturally pierced, some not. If some of the shells they gathered had no hole, how could the hole be integral to the function of the objects? One study went as far as to show that random collections of shells contained pierced and unpierced shells in equal measure. Other studies reached opposing conclusions. I don't think that these studies offer any way to finally discern the true function of the shells; they just pit one statistical approach against another. But whether the holes in the pierced shells fulfilled a function or not, one thing is absolutely clear: the holes are all natural. So these holes in themselves in no way indicate the function of adornment. The Plains Indians used precisely the same types of shells as instruments of sound. Not to make music, but purely as tools, rattles shaken to make a noise and flush out game in certain methods of hunting.

But what about the eagle claws?

I had agreed to put my name to the 2014 collaborative study alongside my Italian colleagues' on the condition that we would come up with a joint theory about the symbolic function of the objects. My idea was that we would talk about negative evidence. A positive piece of evidence would be a manufactured object – a simple man-made hole would make all the difference. A simple hole. But after 150 years of archaeology we haven't found our first Neanderthal hole. Nor the slightest hint that would allow us to claim that these objects were actually worn as adornment. An eagle claw, still enclosed in its horn casing, could be a remarkably useful tool for piercing soft material such as leather. The polished surfaces spotted on some of the Krapina claws could have been caused by them being hand-held at the base. Traces of an animal fibre (tendon?) along with deposits of ochre and coal at the base of one of these claws could indicate that the claw might have been coated at its base by a resin. The use of resin to form tool grips has been widely documented in the Neanderthal context; the technique involved applying fibres along with a mixture of ochre and coal, which provided the glues used with both resistance and elasticity. This interpretation is neither accepted nor even mentioned by the Krapina teams, who again read such clues as backing the flimsy hypothesis of a Neanderthal necklace.

But what about the feathers?

Feathers are usually seen as purely symbolic, as they have no obvious practical function or nutritional value. Their beauty has been exploited by a huge number of traditional societies on earth. It is clearly an objective beauty, not merely a by-product of a subjective view of feathers on the part of certain human cultures. In the natural world one of the functions of large coloured feathers is directly to do with male display to attract females. It is the peacock's tail principle. So aren't humans merely copying birds here? Put in those terms, yes. Neanderthals did decorate their bodies using modes of expression very directly evocative of behaviours recorded in our own *sapiens* populations. So this is objective, universal beauty, crossing not only different cultures but also different animal species. The beauty of a feather is not cultural, it isn't unique to the Navajos or the Kayapo Indians, it has the power to mark and impress all human culture, and more widely all living creatures, from *Homo sapiens* to the capercaillie, taking us back to the ideas of Lucien Scubla and André Leroi-Gourhan, who drew connections between the human and animal worlds: plumage and song in birds, adornment and musical rhythms in humans. But since humans do not naturally grow feathers, the collection and transformation of natural products into cultural decoration would mark a fundamental dialectical shift in the structure of human societies. And if there is doubt over the real function of eagle claws (piercing tools?) or that of shells, none of which were ever deliberately pierced by Neanderthals, systematic research of large feathers does raise much more profound questions, because, instinctively, we do not see them having any other function than a visual one.

Where does that leave our feathers? If I were being cynical, I could point out that feathers can make excellent toothpicks, and we know that Neanderthals used toothpicks by analysing the marks they left on their teeth. But I'm not trying to be cynical. I'm trying to understand. I'm trying to grasp the creature as it was, by putting aside my own feelings, my own desires, my own projections. I'm trying to understand despite myself. To fight my own instincts and open my eyes to the creature itself and to those more distant ancestors, since the gathering of feathers has been recorded more than 420,000 years ago.

Unfortunately, the feather has been prematurely consigned to the category of the useless and the inedible. Instinctively, as Westerners, we find it difficult to imagine what might be edible in a feather. This is probably another example of how we have lost much of the knowledge of hunter-gatherer societies.

Some words from our polar explorer Jean Malaurie spring to mind: 'I was going to go hunting, to drive my sled, to eat raw meat, to suck the pinkish-white and very fragrant energy-giving fat from the stems of the bird feathers and rotten bird bones (the famous Kiviak).' We hit a snag. Not only are these feathers not waste products lacking in proteins, but their proteins possess exceptional energy-enhancing properties and are much sought after by the Inuit.

So the feather too must fall, along with the shells with their crab holes and that pretty bracelet of eagle claws. In our imaginations, our fantasies, our ideas, our projections. Once again, it is *Sapiens* looking at the Neanderthal, *Sapiens* clothing the creature in the garb of *Sapiens*, conceiving it as a version of ourselves, incapable as we are of conceiving a humanity that is not us.

Some might accuse me of applying double standards here in making out that what we interpret as adornment in *Sapiens* should be seen as something entirely different in Neanderthals. But it's a cynical accusation, an outright lie. A refusal to accept, to see, to understand. Why?

Because when we interpret a naturally pierced shell as a type of adornment in *Homo sapiens* it's because we know of millions of shells deliberately pierced and undoubtedly worn as adornment by these humans. We're also forgetting that the oldest shell adornments, discovered in Blombos in South Africa and which are some eighty millennia old, were manufactured and are not simply an assemblage of shells pierced by crabs. The presence of this or that micro-trace of ochre pigment on these shells or on the grips of claws adds nothing to their interpretation. Ochres were used in many different techniques, ranging from the treatment of animal skins to protection from the sun or the concocting of resins to improve the grip on tools. It is fairly ubiquitous in prehistoric soils, where tools and bones are randomly tainted with the powder. The presence of ochres, haematites or other materials which have the collateral effect of coloration gives us no definitive information concerning the aesthetic sensibilities of these populations.

It is likely that a certain number of the pierced shells discovered in Palaeolithic *sapiens* sites and interpreted as adornment in fact had a purely practical function: weights, hunting rattles, tools for stretching fibres or ropes. But in *sapiens* societies millions of shell adornments have been documented, sometimes draped around bodies in their graves. The error then has no particular consequence. The false interpretation does not affect our interpretation of these past populations. But to project this on to a humanity of which we know nothing, and which has not, according to 150 years of our documentation, produced the shadow of a manufactured adornment, would be a major error. An error with possibly disastrous implications for our understanding of these past realities.

Whether drilled into a shell, a tooth or a bone, the first Neanderthal hole has simply not yet been discovered. So we imprison this humanity in our own conceptions, in our own ways of being in the world. This means we refuse to study these populations for what they really were. We make the Neanderthal into a poorly dressed scarecrow; it's not him, it is us. We kill the creature not once but twice.

Armed with this knowledge about these amazing 'Neanderthal symbols', we can now examine the origin of cave art and that 2018 study that made the cover of one of the world's most prestigious scientific journals: 'Neanderthals Made Cave Art . . . Really?'

It doesn't stand up. It isn't consistent with any of the millions of pieces of data we hold on these populations. It is perfectly consistent with a certain purposeful view, which has been emerging in recent decades, but it is not consistent with archaeological fact. After 150 years of archaeology, and millions of cubic metres of Neanderthal soil excavated, where are the ornaments created by craftsmen, or the clear and obvious necklaces? Where are the ivory statuettes? Where are the shale platelets decorated with symbols? Where are the bones engraved with friezes of horses or bison? Such objects were produced in almost industrial quantities in the Palaeolithic *sapiens* societies which were supplanting the Neanderthals at that same time and in that same place.

The Neanderthals were remarkable craftsmen. Why wouldn't they have transformed materials for aesthetic, visual or symbolic purposes? If archaeologists have not yet found any such transformed objects, manufactured objects, it is either because they don't exist or because they are so exceptional that we perhaps haven't encountered them yet, or haven't recognized them yet. If such objects were to emerge from old Neanderthal caves, then their deep significance would have to be evaluated without projecting our own conceptions of the world on to them. Such discoveries would definitely expose us to mental worlds that we have not been able to document hitherto, perhaps also because, fundamentally, we see the Neanderthal as *Sapiens.*

Laden with these questions, we decided, with Jean-Michel Geneste, director of the scientific team at the Chauvet cave, to set up a team to analyse precisely the physical-chemical measurements of the datings that led to the identification of this Neanderthal rock art. Our study was published in the same journal, *Science*, and concluded that the ages obtained in the Spanish caves could not be validated on the bases – including the physical-chemical evidence – presented in the same journal. This Neanderthal cave art was more an article of faith than a matter of hard science.

Doodling and Other Monkey Business

At this stage there is not much left standing of Neanderthal art. Interpretations of confusing marks, parallel or crisscrossing lines on bones, pebbles or on the floor of a cave. Lines. Doodles.

In 1962, the zoologist Desmond Morris published a remarkable book, *The Biology of Art: A Study of the Picture-making Behaviour of the Great Apes and Its Relationship to Human Art.* He reported on the effects on chimps of being taught how to use a piece of chalk or paintbrush. The chimps loved doodling, and each individual developed their own style, which evolved over the years, whether fan-shaped marks or circular or cross-shaped ones. Morris didn't draw parallels between his own research and the prehistoric data, which he was unfamiliar with; it was Franck Bourdier, one of the great prehistorians of the twentieth century, who made potential connections between the experiments with chimps and what archaeology tells us about the Neanderthals.

Sixty years later, and despite frantic efforts to examine the tiniest pieces of evidence left behind by these populations, from a dispassionate point of view the few doodles are the nearest thing we have to this increasingly tenuous Neanderthal 'art'. Could the creature be so far removed from us in its neuronal structures? Is it possible we aren't searching well enough? Or we are searching in the wrong place? Or do our narrow projections on to this humanity prevent us from asking the questions that would allow us to evaluate the objective reality of these populations?

Whatever the reasons for this Neanderthal paradox, neither the few shells nor the feathers nor the claws nor the colourants get us anywhere near defining an aesthetic sensibility on the part of the Neanderthals, and, up until now, it is we, and only we, who have decked the creature out in these coloured pendants. And when you are made up to look like something you are not you tend to look ridiculous.

We will have to reopen this file as soon as we find the first hole, the first slot, the first evidence of a deliberate artisanal transformation of objects with singular shapes and colours. We will need then to ask ourselves what these acts really signified. For now, nothing to see here, the file is empty. Or maybe it is overloaded with our own preconceptions about what humanity means.

There is no sign of a Neanderthal necklace in the 300 millennia that the Middle Palaeolithic lasted. What, in that case, are we to make of the body adornments that emerged in the very last millennia of the Neanderthal era, when the Châtelperronian culture produced the most ancient artfully crafted pendants in western Europe – pierced teeth, carved ivory discs, grooved fossils?

The Fall of Our Last Neanderthal Fantasy.
Neanderthal, Where Art Thou?

As in a novel, the twist in the tale comes near the end, throwing a new light on the whole plot and upsetting the reader's expectations. So it is with the Neanderthals. We hoped for a long time that we would understand them once we were able to see their ultimate artisanal creations, their ultimate tools, the evidence of their genius before their demise. The deciding moment when poker players show their hand and take their chips. For a long time, the quest for a Neanderthal art came up against an almost complete dearth of Neanderthal symbolism. No bodily adornment. No painted caves. No engraved decorations. The quest then focused on the slightest Neanderthal scribbles as evidence of the creature's creative expressiveness, a true pareidolic chimera, like seeing shapes in the clouds, which has never given cause for satisfaction, just as some remarkable data seemed to emerge of the final cultural expressions of these Neanderthal societies, specifically within the Châtelperronian culture.

The Châtelperronian is a prehistoric culture which was apparently limited to certain regions of western Europe. In France it seemed to be restricted to the region between the Pyrenees and Burgundy, excluding the Mediterranean coast and the Rhône valley, which was one of the most important migratory corridors on the European continent. A few isolated outcrops of these astonishing Châtelperronian traditions are to be found on the Iberian peninsula between the Cantabrian corniche and the Mediterranean coast. Such traditions are also temporally specific, replacing the classical Neanderthal traditions in their final millennia, between 40,000 and 45,000 years ago. Scientific consensus over the last seventy years has largely attributed the Châtelperronian to the final Neanderthal populations. This attribution, however, remains hotly debated by some scientists, myself among them, since at closer view the origins of these traditions appear to be only superficially rooted in the Neanderthal knowledge documented in the same geographical locations.

There is good reason to think this: in the Châtelperronian, the heavy splinters of the Mousterian were finally and definitively replaced by slender blades. Shaped into backed points, these relatively standardized objects have very little similarity to earlier Neanderthal artisanal work. They display clear affinities with the tools and weapons of the modern populations that would subsequently occupy Europe. In the Châtelperronian culture, at a point in time when *Homo sapiens* was spreading throughout Europe, the Neanderthals supposedly updated their craft skills in the style of the new world order. How is this possible? By what extraordinary chance did these societies, which had not changed their artisanal traditions for hundreds of millennia, suddenly invent new ways of being that would ultimately be those of their successors, *Homo sapiens*, for the last thirty millennia of prehistory?

In the 1950s, in archaeological excavations of Châtelperronian settlements in the Renne cave at Arcy-sur-Cure in Burgundy, André Leroi-Gourhan uncovered 'modest human remains, but found at every level'. Largely these consisted of teeth whose singular morphology suggested that it was the Neanderthals who were the makers of these amazing Châtelperronian objects. The prehistoric artefacts were remarkable and also included superb jewellery made from teeth and ivory. There were a large number of points made from bone and reindeer antlers and small pierced pearls among the fossils. The artisanal traditions of ancient prehistory that had prevailed over hundreds of millennia had suddenly undergone a revolution, and these Châtelperronian objects display all the modernity of this new world and its new ways of being.

The Neanderthal attribution was supposedly confirmed in 1979 with the discovery of a Neanderthal skeleton on the site of Saint-Césaire in an archaeological unit attributed to the Châtelperronian. The Neanderthals then must have undergone a sudden rupture with their own artisanal traditions, with their own mental universe shaped and renewed from generation to generation, practically since the dawn of time. This new Neanderthal world, with its profusion of objects manufactured from animal materials, showed a mastery of the functions of signs, symbolic statements and individual and group aesthetics.

So just before the end, on the very eve of their extinction, the Neanderthals dropped their mask. They were fully human, and their technological stasis over more than 300 millennia had been simply the expression of a strange cultural choice not to display their deeply modern nature.

Given the improbable convergence between this creative explosion and the arrival of *Homo sapiens* in Europe, some naysayers have suggested that the Neanderthals were merely poor imitators of the modern humans who had just arrived on their territory. The creature copied without understanding, merely parroted phrases without understanding the grammar. And yet the different archaeological horizons of the Châtelperronian at Arcy arguably revealed more objects overinvested with meaning than all the archaeological evidence of the first modern humans on the European continent put together. We are left with a rather strange choice between, on the one hand, a Neanderthal genius and, on the other, a mere imitator of the new modern arts: two opposing views. This is not a battle between the Neanderthal and modern humans. It is our battle. It divides the scientific community into two irreconcilable camps. The Neanderthal no longer exists, except in our minds. There is no simple solution to this, either in our sciences or in our minds.

Just as the question seemed to have been definitively settled for more than thirty years, a 2010 study by the Oxford laboratory threw a spanner in the works. It presented a detailed analysis of the Châtelperronian layers at Arcy-sur-Cure based on a large corpus of carbon-14 datings. The results were to cause quite a stir. The ages at Châtelperronian Arcy spanned more than twenty millennia between 25,000 and 45,000 years ago. However, the Châtelperronian traditions lasted only four to five millennia. The excavation had been meticulous but had used ancient techniques and had failed to order the archaeological remains by age and jumbled together radically different objects from different times. The extent of this muddle, trivial for some, fatal for others, has been and remains the subject of bloody controversies in which two scientific camps have taken up entrenched positions, without much prospect of compromise or real dialogue.

Another archaeological site might throw some light on this question. It is 1979, and we are in Saint-Césaire, a small town in Charente-Maritime, near Saintes. The trowel of François Lévêque encounters the remains of a Neanderthal body. The discovery was sensational: Neanderthal remains are rare – in fact, since then, no other body has been discovered in France. Just as importantly, the body is found in a Châtelperronian layer, which seems to close the debate about where these first modern traditions came from. But almost forty years later, in 2018, a team from the University of Bordeaux, led by Brad Gravina and Jean-Guillaume Bordes, would question the link made between the body and the remains by demonstrating that the Neanderthal body could have belonged to much more ancient settlements and thus tells us nothing about the originators of the Châtelperronian.

Once again, the sequence from the Renne cave at Arcy-sur-Cure was the only evidence for identifying the population that originated one of the first modern cultures of western Europe. But we have seen that all the archaeological layers at Arcy remain matters of lively debate, and, given the dearth of any new discoveries, the smallest bones at Arcy are being made to talk, counterbalancing the problems of the age and homogeneity of the collections with a host of biomolecular analyses and direct datings. But this is like extracting a confession under torture. We have seen how unreliable radiometric dating is for this period, and biomolecular analyses are based on the assumption that there is a clear distinction within the genus *Homo* between *Sapiens* and Neanderthal, based on an analysis of the ancient proteins contained in their bones. To this day, the debate is far from being concluded, and although the teeth discovered in the archaeological layers at Arcy are undoubtedly Neanderthal, their association with the Châtelperronian artefacts cannot be firmly verified. Analysis of the artefacts at Arcy actually revealed surprising associations between modern and classically Neanderthal technologies. Rather as if we discovered a transistor in a Roman villa.

Since the 1970s, then, there has been no new complete sequence, with flints, bones and human remains, which would enable us to update our knowledge about the precise structure of these societies, one of the first modern cultures identified in France. So who was the originator? *Sapiens* or Neanderthal?

The very few Châtelperronian sites identified in these last forty years have yielded only paltry quantities of bone remains. Without bone remains, how can we know these populations? Their hunts, their logistical strategies and their organization are no longer directly available to us. Just a handful of bones over forty years, degraded by soil and time, are all we have to show from the few new Châtelperronian sites, and this impacts directly on our precise understanding of these societies. As for the question of whether it was the Neanderthals or *Sapiens* who initiated these traditions, if we exclude the skeleton of Saint-Césaire, which we cannot associate with the Châtelperronian with any confidence, we basically have to rely on remains found at a single site and extracted from the soil more than sixty years ago. Since then, not the smallest verified human bone, not a single tooth. No bone tools, no ivory or antler adornments at the very point when these artefacts were first appearing. This is damaging, to say the least, because these are precisely the artefacts that mark the entry of human societies into recent prehistory, marked in Europe by the transition from the reign of Neanderthals to that of *Sapiens*.

The modernity of the Châtelperronian could well be the work of the Neanderthal. But it is also possible that the vestiges of this transistor are not Roman and that by focusing on the electronic components we miss the point and force the creature rather prematurely into our own image.

The expanding field of palaeogenetics may in the end demonstrate that the originator of the Châtelperronian is not the one we hoped for. Is it possible, then, that this culture does not represent the last gasp of certain Neanderthal societies but rather marks the arrival of *sapiens* populations in European lands?

I am not presupposing any answer to this question. Keeping an open mind, I recently published a paper deciphering the striking relationships that could be woven between the structures of the Châtelperronian and those of some contemporary traditions documented in the eastern Mediterranean. In this study I identified the Levant as the geographical location where this culture emerged. But here on the sides of Mount Lebanon such technological traditions are undoubtedly attributable to modern humans.

This hypothesis invites us to totally redefine our ideas about the first modern traditions in western Europe and attribute the Châtelperronian exclusively to *Homo sapiens*. As it stands, it seems important that we factor in doubt to our approaches. Should future research confirm my hypothesis, and in view of the ambiguity of the evidence as to the existence of a Neanderthal art, then there is currently no robust scientific basis for the claim that Neanderthals and *Sapiens* followed convergent evolutionary trajectories leading with surprising synchronicity to a similar emergence of symbolic thought.

This chapter may have come as a surprise to many, both specialists and those with a general interest in the subject, given how major scientific journals and the mass media more generally have framed the Neanderthals in a narrow anthropocentric vision. In my view this is at odds with archaeological facts. It refers to no obvious reality of Neanderthal material or non-material production and tells us nothing about the exact mental structures of these peoples.

So my reading of this 'Neanderthal volatile art' is not simply an exercise of deconstruction. It is not intended to provoke, or to gratuitously question the nature of the creature. It refocuses the debate and tackles their material reality head-on. It asks questions about notions of the historical and ethological truth of populations of which we know little, and which we urgently need to stop dressing up as ourselves.

The Badly Dressed Scarecrow

You've probably heard it said that you wouldn't recognize a Neanderthal if you met one on the subway? Well, it's not true.

You will find umpteen variations on that theme in various articles and interviews about the creature. Researchers generally attribute this canard to the writings of William Strauss and Alexander Cave, published in 1957 to mark the centenary of the discovery of the Neanderthal specimen type. But in fact it originated nearly twenty years earlier in the writings of Professor Carleton S. Coon. In his 1939 work on the races of Europe,* Coon used this image of a clean-shaven Neanderthal in a suit and tie and a trilby to 'illustrate the fact that our impressions of racial differences between groups of mankind are often largely influenced by modes of hair dressing, the presence or absence of a beard and clothing'. This was written at a time when ideas of classifying human populations would lead the West into a

---

* Carleton Coon, *The Races of Europe*, Macmillan, 1939 (update of a work by William Z. Ripley of 1899).

catastrophic world war. In this same year, 1939, Coon discovered a fragment of a jawbone in a cave near Tangiers in Morocco which he identified as Neanderthal – today we know that it in fact belonged to a *Homo sapiens*.

Coon's work divided the peoples of Europe into seventeen races, some of them cross-bred with Neanderthals, which he contrasted with the 'pure *Sapiens*' of Mediterranean Europe. The image of this Neanderthal dressed like a New Yorker in his trilby hat wasn't Coon's way of rehabilitating the Neanderthal but was part of a more sulphurous ideology that envisaged continuities of population and of morphological traits in different parts of the world.

Even if this Neanderthal on the subway was not in any way the beginning of the creature's rehabilitation, it was nonetheless a purely imaginary construction. You will note that the hat

Illustration published in 1939 by the anthropologist Carleton S. Coon in his book *Races of Europe*.*

* The Macmillan Company, cf. https://archive.org/details/in.ernet.dli.2015. 222580/page/n5/mode/2up, p. 24.

hides the most salient morphological features, which are located in the upper part of the head, from the eye sockets to the back of the skull – supraorbital torus, receding forehead, occipital bun. We have to concede that, hidden behind fabrics, veils and disguises, no one would be able to recognize a Neanderthal on the subway. But now we are in the land of Grimms' fairy tales, where the wolf mimics his own victim: 'The grandmother was asleep, most of her face covered by her bonnet, and she looked so strange.' And remember that both the wolf and the Neanderthal came to a sticky end.

This image of the Neanderthal in a suit travelling incognito on the subway has kept cropping up in public talks and exhibitions since the 1930s. He is disguised as one of us, a product of our fantasies and a remarkable symbol of our Western projections. This image has been put forth by successive generations of scientists. This Neanderthal on the subway is in no way a rehabilitation of the creature based on the most recent scientific discoveries. He is part of our myths, our representations. It is a purely ideological creation that tells us nothing about the Neanderthals, but plenty about our own societies, our own taboos and intellectual puritanism about the very concept of difference.

We can see what is at stake here in the mere representation of the Neanderthals. The bone structures that characterize them are in fact intrinsically only of secondary importance in understanding who these people were. The shape of their skulls would be rendered more or less irrelevant if the Neanderthals had in fact been able to merge into what we are at this point in time, *Homo suit-and-tie*. But the shape of the skull is only ever an empty shell, the easy part, morphologically classifiable and analysable even on a basic computer. The 'deep' Neanderthal, the real being, has never been fossilized in any of the bone structures of its skull. Understanding that intangible matter inside the shell of the skull requires an intimate acquaintance with the creature, who remains an unknown quantity to anthropologists or geneticists. The structure of thought is by its very nature subtle, subject to interpretation; it can only rest on the most intimate possible knowledge of the traces left behind by these people.

This is about their technical knowledge, their way of applying it, their relationship with the natural world and the mineral world, their perception of the world of the living and the world of the dead, their ideas about themselves and others. That is where it is all played out. And that is why, unconsciously, we constantly project ourselves in this representation of the Neanderthal in a suit and tie. Because clothes maketh the man. Of course they confer – impose – humanity on the creature, but this humanity is not ours. In every human society, whether a traditional one or one that is thoroughly globalized, codes of dress establish social rank as well as human status. By this I mean that, without going so far as to determine structural differences in mental abilities that may well distinguish us from the Neanderthals, simple cultural differences, buried in our unconscious, distinguish the human from the *infra*-human.

There is an abundance of ethnographic literature on representation through dress and the notion of humanity. Saint-Exupéry reminds us that the West is a society without self-reflection, trapped in its ethnic codes:

I have good reason to believe that the planet from which the little prince came is the asteroid known as B 612. This asteroid has only once been seen through a telescope by a Turkish astronomer, in 1909. At the same time, this astronomer made a grand presentation of his discovery before an International

Congress of Astronomy. But since he was wearing Turkish national costume, no one would believe him. Grown-ups are like that . . . Fortunately for the reputation of Asteroid B 612, a Turkish dictator ordered his subjects, on pain of death, to convert to European dress. In 1920, our astronomer repeated his demonstration, wearing elegant evening dress. This time everyone accepted his proofs.*

Saint-Exupéry points to the structures of self-representation in his dazzling novel, written in 1942, at a time when the great and good, prisoners of their own history, would show their most detestable faces. He shows the absurdity of judging individuals and populations on the basis of their appearance at exactly the same time that our Neanderthal appears in the subway, rehabilitated, virtually disguised as us. Under the pretence

Antoine de Saint-Exupéry, *Le Petit Prince*.†

* Antoine de Saint-Exupéry, *The Little Prince*, trans. T. V. F. Cuffe, Penguin, 1995, pp. 15–16.
† Copyright by Gallimard, 1943.

of rehabilitation, this travestying of an individual by means of clothes and postures that are not part of their culture is actually assimilation. By looking at it this way we can perceive, under the superficial and self-righteous layer of our projections, much more insidious realities that relate to some of western history's problematic moments. This Neanderthal in the subway calls to mind the Cultural Assimilation of Native Americans, launched under the aegis of George Washington and Henry Knox, by which the USA between 1790 and 1920 engaged in a process of enforced cultural transformation of Native American populations. In 1879, the Carlisle Indian School, founded by Captain Richard Henry Pratt, had as its motto: 'Kill the Indian . . . and Save the Man'. Students at the school were compelled to cut off their hair, give up their language, their traditions and their traditional clothes and instead speak English and dress like Americans, a policy that was still in place in the USA just a decade or so before the appearance of our portrait of the Neanderthal in his suit and tie.

Dressing up the Neanderthal in a trilby is a risky game. When you play with fire you end up getting burned. Not because this headgear hides a rather low forehead – bone structure tells us nothing about mental capacity. But presenting the creature in this way and, what is more, continuing to do so nearly a century later, leaves no possibility of understanding what the Neanderthal really was. The approach now essentially consists in playing on the unconscious of the general public, who at best have only a limited knowledge of these populations, since they are taught nothing about them at school, in order to imprint an image in their minds that brooks no discussion. Neanderthal? The same as us. Full stop. And this is at once a prejudice and, scientifically speaking, a lie. This lie imprisons our imaginations, our understanding of past

Tom Torlino, a Navajo, at the start of his time at the Carlisle Indian School in 1882 and how he looked three years later.*

realities. It is not just a lie for the general public: it can directly affect the scientific community too, through this strange game of media-hyped discoveries that systematically aim to demonstrate, prove and close the debate on the categorically human nature (= us, our equal, our mirror) of this population. Is the creature then in need of our concepts to rehabilitate itself? To raise itself to the level we are at, or flatter ourselves that we are at? When in 1929, in the writings and ideas of many

* National Archives and Records Administration, RG 75, Series 1327, box 18, folder 872, http://carlisleindian.dickinson.edu/student_files/tom-torlino-student-file.

researchers, we had already shaped it according to a certain vision of ourselves, *Homo suit-and-tie*. The Neanderthal becomes little more than a macabre puppet in the hands of sorcerer's apprentices.

We should bear in mind here that the native populations of the Americas differed from western populations only in their technological and cultural traditions. The Neanderthal is, however, a fossil on three scores: culturally, biologically and ethologically. The question is then to determine whether its biology induced behaviours – an ethology – that were unique to it. And since 99.9 per cent of its artefacts that have been preserved consist of hard stone tools – flint, quartz, obsidian and quartzite – these are the only objects that can speak to us about the mental structures of these populations. But what do these objects tell us? And what do they suggest about the other humanities that were its contemporaries? We obviously can't limit this to technical competence, for the simple reason that it is purely technical, and our humanity defines itself and distinguishes itself through the soul that goes into the work, by the investment of our objects with cultural, symbolic and transcendent values. Our objects are haunted by our signs, by our irrationality.

We all know that instinctively; if our objects weren't above all the expressions of our fantasies, then a Van Gogh would be nothing but splashes of colour on a stretched canvas. All these irrational forms of adornment, of art, of soul projected into matter are easily recognizable from an archaeological point of view and documented in huge quantities from the very earliest *Homo sapiens* societies and throughout this whole period of the recent European Palaeolithic in the very territories that had been abandoned by the Neanderthals and taken over by this new *sapiens* humanity. These irrational forms, those of our *sapiens* ancestors, are seen in the dozens of worked pearls, human and animal statuettes, flutes, painted caves, figurative and abstract representations using all sorts of materials: ivory, bone, stones, cave walls.

So we have sought projections of ourselves in the Neanderthals. The smallest piece of bone that has been slightly scraped or skinned has been subjected to thoroughgoing investigation. The slightest line has been interpreted for signs of soul. Spirit, where art thou?

6.

Understanding the Human Creature

The Neanderthal was never just another version of us, but we are a long way off understanding its own ways of being human. It's a fascinating process which requires us to fundamentally explore what we – *Homo sapiens* – are ourselves. To explore our nature not as humanity but as *a* humanity.

On Self-consciousness

In April 2021, *Nature: Molecular Psychiatry* published a study that aimed to decipher the emergence of human creativity by focusing on three main aspects of personality: emotional reactivity, self-control and self-consciousness. It revealed the existence of genetic structures in Neanderthals similar to those identified in chimpanzees when it came to emotional reactivity, and a position halfway between chimpanzees and modern humans when it came to self-control and self-consciousness, which directly impacted on their creative potential, their consciousness of self and their prosocial behaviour.

Make no mistake, this will do nothing to change the view of researchers intent on simplifying Neanderthals into another version of ourselves – and their position is not exclusively a dogmatic one. Establishing what is unique about humans is no easier by examination of their molecular components than it was in Antiquity when Plato defined humans as bipeds without feathers. And if these studies look impressively scientific, bear in mind that neither geneticists nor specialists in physical anthropology are equipped to deal with the social, mental, ethological and cultural structures of these extinct societies.

This study offers both a reminder and a serious warning. The reminder is that these three hominins shared a common ancestor: prior to divergence, around ten million years ago, there existed a hominin that was ancestor to both humanities and apes. That time span is only twenty times the one that separates us from the Neanderthals, who lived around half a million years ago. Starting from the postulate that 500 millennia of divergence between Neanderthals and us have had no impact on the neuronal structures of our two populations or that these populations have evolved independently towards the same purpose (another us) is, strictly speaking, a new form of creationism. A more palatable form of creationism, creationism 2.0. Genetics show that there is a thick wedge of time between us and the Neanderthals. If ten million years are enough to separate us from the chimpanzee, this half-million-year period should not be underestimated. This is why we should take this study as a warning.

We have to go back to the archaeological material and interrogate it. Go back to these ancient artefacts, these objects discarded by an extinct humanity. Question them. Question also the encounter between our ancestors and these distant populations. But there is no proof that any encounter between two distinct humanities on the same territory ever took place, and there is no archaeological trace of it. We can only deduce it from genetic information that shows that intermixing took place between populations, but no archaeological site documents these strange meetings between two humanities. They are too distant from us in time, and the elements preserved are too evanescent and too rare to act as proof of such encounters. That is regrettable, as these moments of encounter between different humanoids constitute fundamental episodes in the history of humanity, probably the major turning points in the spectacular development of our species on this planet. It was such moments that structured our humanity's expansion across the ancient world. These moments probably also provided the fundamental keys to the development of the technologies and modes of representation of our ancestors. Lévi-Strauss tells us that human societies are driven by the necessity to differentiate themselves, to mark themselves out, to express their particularity in the organization of their groups. It is not so much the isolation as the proximity of human groups that motivates the need to express what it is that makes us human, to visually communicate our singularity and the reason why 'those others' are not fully human. This need to differentiate oneself from the other is a classic of ethnography, but in this case it occurs between two distinct humanities. The social and cultural implications of such encounters must have fundamentally impacted all of these populations' modes of representation.

In the rare European sites which have been able to preserve remains of both the last Neanderthals and the first modern humans, how much time elapsed exactly between the two settlements? That is the million-dollar question. The Mandrin cave up above the Rhône valley, in which I have been working since 1998, opens a window on this singular moment of change. One day we turned up two objects: a Mousterian point carved out of black flint – a classic example of Neanderthal workmanship – and a small white blade, which analysis showed to be undoubtedly the work of a modern human. These two objects were right next to each other in the ground. Here was an image of physical contact – an interesting piece of evidence but one not easy to interpret. Did the two populations inhabit the cave at the same time? Or did this Neanderthal point lie in the ground for a millennium before the *Sapiens* dropped its blade there? This is basically just a matter of anecdote, so how do we address a question of this importance if the physical proximity of two objects so representative of the two societies teaches us nothing at all about a possible encounter between these populations?

If we rely on information derived from bone traces and physical-chemical analyses we find ourselves in the same dead end. Carbon-14 datings are accurate only to within a few centuries or millennia and so cannot be used to prove simultaneous habitation in the same territory. The contemporaneity of these populations in Europe, as demonstrated on the basis of analyses of the ages of the last Neanderthal occupations and the most ancient traces of *Homo sapiens*, arises from a basic statistical probability, not from any historical or ethnographic reality. In Europe, any meeting between these populations is invisible to us and may never have taken place. The scale of our ignorance concerning one of the most significant events in the history of humanity is staggering.

Fire Memory

Faced with these dead ends, which are not specific to my research in the Mandrin cave but impede the development of knowledge across the whole scientific community, we were fortunate enough at Mandrin to be able to develop a remarkable method based on analysis of fragments from the walls of the cave. This involved analysing soot deposits left on the vault of the cavity when the prehistoric people made their fires there. The study drew on doctoral research by Ségolène Vandevelde and revealed traces of successive generations of Palaeolithic hunters. The deposits of soot on the walls of the Mandrin have a recognizable signature which allows us in this case to clearly distinguish what came from Neanderthal fires and what from *sapiens* fires. This veritable 'fire memory' became our archive. We worked for nearly fifteen years in order to obtain samples from the vault documenting all the prehistoric settlements over the span of eight millennia.

This high-resolution detective work produced an unexpected discovery: analysis of the films of soot revealed that the two humanities inhabited this cave no more than a year apart. A maximum of a year. That meant that, for the first time in Europe, we had evidence pointing to a physical encounter between them in a well-defined territory. The two humanities must have physically met right here. We are unable as yet to achieve a resolution greater than a year, but we have shown for the first time that the two human groups were effectively contemporaries in a very precise territory, whether their encounter took place in the wider territory, the mid Rhône valley, or in this very cave itself. As the Mandrin cave had been continually used for nearly 80,000 years by Neanderthal populations, the fact that the moment of this meeting also marks the end of Neanderthal societies everywhere in Europe can hardly be put down to an unfortunate coincidence. Not only do we find no more traces of Neanderthal cultures after the exact moment of the encounter, but it seems these populations ceased to exist, biologically speaking, outside of a few peripheral areas of the continent, which takes us back to the possible polar zones of refuge that we discussed earlier.

And since, contrary to a number of my colleagues, I totally exclude the possibility that the Neanderthals died of a blast of cold or that they evaporated like snow in the sun, then I naturally deduce that the reasons why the Neanderthal populations disappeared were fundamentally linked to the arrival of this other humanity. Whatever the nature of their relationship, at Mandrin and elsewhere, new populations of evidently very dynamic modern humans moved into these territories and replaced the aboriginal Neanderthal populations that had been established there for tens of thousands of years.

We can assume that this wasn't just a progressive displacement of the Neanderthals as *sapiens* populations slowly migrated west over the course of centuries or millennia but a fully fledged conquest. The archaeological records indicate multiple waves of population. I think there were three very distinct waves, with the first two failing to definitively occupy these territories. The third wave was a true culturally homogeneous population wave which very rapidly occupied the whole of the continent. These were the early Aurignacians, whose descendants would paint the Chauvet cave in the same region of the Rhône valley. The Neanderthals gave way and did not make a comeback, their line extinguished, everywhere, around the forty-second millennium. And here, in the Mandrin cave, we know that the replacement happened over the course of a few seasons. In other words, the Neanderthal populations were ousted in the most abrupt manner. It might be argued that this short time frame is just the product of the low resolution of our methods, such as carbon 14, which is accurate only to a few centuries or millennia. Chronologies based on genetic analysis of fossil populations do not offer any better resolution; in fact, despite appearing to be founded on hard science, they might even be more uncertain. But our soot, our fire memory, has transcended such methodological limitations. Here, the population replacement did not take place over a few millennia, or a few centuries, or a human lifespan, it happened in the blink of an eye.

So the Neanderthals didn't just evaporate and they didn't merge genetically with us. They weren't wiped out by a comet or a volcanic eruption. Nor did they become sterile after 300 millennia of existence at the point, the very year, when *Sapiens* arrived in their territory. We can draw other parallels: although the native peoples of the Americas were effectively decimated by viruses and bacteria brought over by the Europeans, none of these populations was wiped out by smallpox, measles, typhus or cholera. The eradication of these aboriginal American societies was primarily attributable to the Europeans and a chain of events, historically distinct but nonetheless consistent with the influx of a colonizing population.

'I take your sister but I don't give you mine'

Let us return to the theories that baldly contend that these fossil humanities did not become extinct in whole or in part but instead merged biologically, genetically, with us. Genetics in fact has told us nothing about the fate of the last Neanderthals, since the small percentages that survive in current populations seem to derive from much earlier interbreeding, perhaps around the 100th millennium, somewhere in Asia. As far as Europe is concerned, when we are able to reconstitute some of their genetic information, we find that the first *Homo sapiens* in our latitudes systematically possess Neanderthal ancestors. This was established on the basis of bones discovered in Romania, Bulgaria, the Czech Republic and Siberia.

However, palaeogenetics shows no mixing between *Sapiens* and the last Neanderthal populations during the phases of colonization. In other words, we have not discovered a creole Neanderthal population created from a hybridization between Neanderthal and *Sapiens* at the moment of this human extinction. The genetic exchanges thus seem to have functioned in one direction only: from Neanderthal to *Sapiens*.

This paradox contains arguably one of the most important pieces of information about the relations that existed between the two populations. We find, in this contact phase between populations on the European continent, not only that the presence of Neanderthal or Denisovan DNA is well documented among the first *sapiens* populations, but that gene flows between different populations of the Pleistocene appear to have been systematic. However, surprisingly, the opposite is not true, and the genetic sequencing of the most recent Neanderthal populations in Europe shows the absence of any *sapiens* ingress into aboriginal Neanderthal populations. And the development of palaeogenetic analyses seems to confirm a similar pattern in every new genetic sequencing of old *sapiens* populations. The implications for relations between Neanderthals and *Sapiens* appear to be crucial and may well provide the first global understanding of historical and ethnographical interactions between the two populations at this moment of colonization in Europe.

Such a paradox may hold a key to the types of relations that existed between these populations at the time when *Sapiens* was expanding in the far west of Eurasia. We know from the work of Claude Lévi-Strauss in 1949 on the elementary structures of kinship that the exchange of women is a fundamental, invariant feature of the organization of every human society. By way of alliance between two human groups women are systematically integrated into the group of men. Genetics suggests that this 'patrilocality' was already practised by Neanderthals. But this exchange of women, which ensures the biological survival of the population, is based on reciprocity: 'I give you my sister, you give me your sister.' Aside from ensuring the simple genetic survival of the two groups, this act creates or enables an alliance between the two peoples. The absence of signs of *sapiens* interbreeding in the last Neanderthals and, conversely, their widespread presence among the first *Sapiens* in Europe could then represent a fundamental indicator as to the nature of the relationships between these populations, whether they took place in Europe or in Asia.

Palaeogenetics, then, reveals an unexpected non-reciprocity which might be summed up as follows: 'I take your sister but I don't give you mine.' This lack of reciprocity in one of the fundamental structures of the relations between populations is disturbing. In ethnography the exchange of genes is not about love but is rather foundational and characteristic of the structure of alliances between human societies. If future palaeogenetic analyses confirm this systematically asymmetrical arrangement, we may well have here the first key to a robust reading of the rather uninspiring relations between these populations at the time of their encounter in Europe. It would mark out a path to understanding the process of extinction of the Neanderthal populations while illustrating the origin of certain Neanderthal genes in modern-day populations of Eurasia. Unfortunately palaeogenetics has not yet told us enough to develop these pathways further, but for the first time we will perhaps be able to glimpse the real nature of the interactions between the Neanderthals and the *Sapiens* who colonized the European continent.

Tracking the Neanderthal from Valley to Valley

Palaeontology does not document exactly what interactions took place between the two populations at this moment of disappearance. But the fact is, once *Sapiens* started to turn up in our archaeological records, Neanderthal disappeared. This is an archaeological given, staring us in the face, with an obvious relationship of cause and effect that repeated itself across every region of Europe until the species was completely extinct.

It is worth underlining this again: the Neanderthals are an extinct population. Completely extinct. Let's say all populations of wolves were to disappear; if we tried to relativize this extinction by saying that poodles or chow chows or shar-peis still bear most of the wolf's genes, the argument simply wouldn't stand up. The Neanderthal is dead, and Granny's poodle is not a wolf, luckily for Granny. And paradoxically it seems to be those dogs that bear the least resemblance to the wolf that are closest to it in genetic terms.

As for the areas where Neanderthals survived beyond 42,000 years, for example the Byzovaya site on the flanks of the polar Urals – where I discovered Mousterian artefacts dating from later than 28,500 years, but no bones that would identify the maker – it seems unlikely that these were populations that were ejected from the heart of Europe. Such Arctic societies were probably native peoples who simply continued to live in the old way and whose traditions were not extinguished until a few millennia later. And we know nothing about the eventual fate of these boreal societies. We can argue endlessly about dates, climate, or this or that other factor, but we have to accept that we are faced with a clear and radical replacement of population. Again, when *Sapiens* made an appearance, Neanderthal disappeared from the archaeological record.

So the implication is that the Neanderthals did not die a beautiful death. However, there is no archaeologically documented evidence of any conflict. To find archaeological traces of a Neanderthal–*Sapiens* war we would need to uncover many more archaeological sites from this key period. We have only very few excavations in Europe that include well-documented Neanderthal occupations, say between 40,000 and 44,000 years ago, and where the reliability of dates is beyond question, such as at the Mandrin cave. And the probability of any of these sites being the exact place where these populations confronted each other is by definition minute. An archaeological search is just a tiny window on to the distant past. We can't look around to see what happened in the surrounding area. For that we would need well-preserved archaeological layers that extended outside the cave and research methods for finding them. It would require not just an ensemble that is exceptionally well preserved but also a different kind of archaeology to excavate it. To dig not just inside a cave, even though that would remain a strategic area of research, but on a hillside, in a valley, to track our Neanderthals via neighbouring valleys, follow their occupations from valley to valley, from cave to cave. There may well be remains preserved on this scale, and it is possible to acquire the means for such research, which would be an exemplary scientific programme, but, to my knowledge, nothing so ambitious has ever been developed.

Mandrin is a small cavity, but generally speaking a cave is not where these populations led their lives. It is a remarkable place, as the vault creates a singularly welcoming shelter, but the natural locus of life of these populations at the time of their settlement at Mandrin extended outwards into a space without boundaries or limits in the sense that we understand today. We cannot imagine, given the sedentary lifestyles we have inherited, in our artificially demarcated world of walls, barriers, roads, grids, the sense of freedom of nomadic hunters living in an absolutely natural world in which humans formed only a marginal presence. We open sites that are only a few dozen square metres, a tiny window on to the past, a window that has no correspondence with the real spaces inhabited by these groups. If conflicts took place 2 kilometres or even just 10 metres outside our zone of research, we simply would not find the slightest trace. And we would probably end up knowing nothing of this history.

It is conceivable that no traces of conflict have been preserved from such distant periods. We cannot compare our ancient archaeological sources with evidence of conflicts at the very end of prehistory, during the Neolithic, as this evidence concerns periods that are exactly ten times more recent. We have found evidence of slaughter in the Neolithic for the simple reason that these events took place only a few millennia ago. The Neanderthals disappeared in a vastly more distant time period. Compared to more recent periods, we have very few sites and hardly any Neanderthal bodies in anything like a complete state. We have discovered about forty, covering a period of 300 millennia. Absence of conflict simply boils down to an absence of the bodies of these last Neanderthals. To state that there were no conflicts because we haven't found any archaeological traces is as valid as saying that the Neanderthals have not disappeared because we haven't found their bodies. What we have before us is simply an archaeological void, a lack of visibility, not a proof that there were no conflicts.

So there was one year at the most separating our final Neanderthal populations and our first *Sapiens*. Here, but no doubt elsewhere too, an encounter must have taken place. As for the extinction, and its synchronicity, that now seems beyond doubt given the exemplary record of the Mandrin cave. So we have the scientific evidence to enable us to analyse this human extinction. We have the physical encounter, the sudden replacement and the completion of the process: the extinction of the species.

Let's say that we found a group of bodies lying side by side in the Mandrin cave; it would be a delicate business proving that this was a massacre. Identifying traces resulting from acts of violence is dependent on how well preserved the bones are. It is worth stressing too that sites that do contain Neanderthal bodies generally deliver up several at once. Ten bodies were discovered at the Shanidar excavation in Iraq – out of the total of around forty Neanderthal bodies identified over the last 150 years of research. This single ensemble alone accounts for a quarter of the main data on the Neanderthals. That would mean either that the cave specifically functioned as a burial place over the course of time or simply that this immense cavern, a prominent feature on the landscape, would have attracted a large Neanderthal population over the millennia, which would make it statistically likely that a number of bodies would have accumulated there. The presence of multiple bodies in one site or even a single archaeological layer can never be interpreted easily, since a layer can record human activity over many centuries or even millennia. Thanks to our analysis of the soot at the Mandrin cave we know that Neanderthals occupied the cave more than a hundred times and apparently never left any bodies there. Was that because no one died during those hundred or so occupations? Or because they would not have buried their dead in this cavity since that was not part of their customs? Or because the bodies were placed elsewhere, outside the cave, perhaps quite near to the area we searched?

An Echo across the Millennia

In any event, there is a before and an after around the forty-second millennium. The Mandrin cave is in the Rhône valley, an important migratory corridor which has been central to the European circulatory system since time immemorial. This is not some zone on the periphery; it lies at the very heart of a major artery of continent-wide exchange. So it is not the most logical place to find the last Neanderthals or evidence of the end of their cultures. Quite the contrary, this is where we might potentially document the very first arrivals of modern humans, and since we have posited a causal relationship between the arrival of one population and the disappearance of the other, there seems to be little hope of finding late traces of Neanderthals in this place, unless we can come up with other, more complex processes that might have led to the extinction of the species.

Where can we find the trace of these last Neanderthals, or the diagnostic traces of their artefacts? Perhaps in the south of Spain or on the Arctic Circle, which brings us back to the bend of the Pechora river and those amazing mammoth hunters of Byzovaya on the edge of the polar Urals.

Despite everything, in Mandrin, on a major traffic route, the Rhône corridor, we do find precisely such testimony of the last Neanderthal societies and the first modern humans. Here a distinction could be made between the nerve centres of circulation and satellite geographical zones where these societies survived to a limited degree. At Mandrin as elsewhere, we posit that the disappearance of the Neanderthals occurred in tandem with the advance of *Sapiens*. There is no real gap between the last expressions of the Mousterian and the appearance of the recent Palaeolithic in the various regions of Eurasia. These events simply took place earlier in the Rhône corridor and later in areas more remote from the migratory arteries. The same pattern of replacement repeats itself. This would affirm the hypothesis that the appearance of modern humans was not just the principal factor but, unfortunately, the direct and unique cause of the extinction of Neanderthal populations and their traditional knowledge. By obfuscating questions to do with this human extinction by reference to climate change, genetic and demographic weakness, disease, we might be merely avoiding the implications of a potentially disturbing major historical event. Denying our colonialist guilt, so to speak.

We can't of course extrapolate a global explanatory schema from a single archaeological dig, but Mandrin is a remarkable site, and thirty years of continual investigation have produced robust, original data and a temporal exactitude that is both unique and unexpected; it's a textbook case. To illustrate what we have discovered at Mandrin, it is useful to draw a parallel with the colonization of the American continent by Europeans, and their encounter with a very great diversity of aboriginal societies who had occupied this territory for millennia. Obviously we can't treat all the events that took place then as a single whole. The history of the Canadian north is not the same as that of the Amazon or Tierra del Fuego. Yet, in a historical perspective, these processes seem similar in their outcomes: a replacement of local populations, which impacted on their traditional ancestral skills, their social organization, their values, their ways of life and their languages. It was the general structure of such populations that was eradicated, on two vast continents and in a matter of a few centuries, while in the case of the Neanderthals a whole human species was involved. This history of recent colonization shows us that even if each region had its own history, the general process was nonetheless the same for all the human societies of the Americas. Over several generations, but in a systematic manner, it was the arrival of the European colonizers which destabilized these aboriginal societies and caused their disappearance or, more precisely, their eradication. By examining the dazzling speed at which such processes took place we gain insight into the enigma of the Neanderthal extinction. Was the colonization of the Americas a distant echo across the millennia of the colonization of western Eurasia?

This type of process seems to me scientifically and powerfully indicated in the archaeological facts, and I would be quite surprised if it does not end up becoming self-evident, even if the radical process of human replacement has yet to be documented in detail, region by region. We would need a Mandrin cave every 500 kilometres to fully examine the historical structures of this human replacement, but I don't think that we can remove *Sapiens* from the equation, to which they are logically integral. This evidence, however, is routinely qualified and diminished by some of my colleagues. Some make the objection that the Neanderthal groups had become very reduced in numbers, which led to their genetic collapse; others point to a slight climate warming, causing the growth of a new forest cover, further isolating these human groups and ringing the death knell for the entire species. There are a number of theories like this that could potentially add nuance to the replacement theory. I personally would not add nuance to it at all. The Neanderthal populations did not seem at all dependent on climate or environment: they had freed themselves from such concerns a long time earlier. I don't think there are fundamentally any climatic or environmental factors in the broadest sense that could explain the radical disappearance of a humanity in its entirety. The synchronous arrival of modern humans in Neanderthal territories was the objective structural event in this process, the only one that can comprehensively account for the extinction of this population. We cannot exonerate *Sapiens*. And as for regional variations across Europe, we can point to the wide variety of situations encountered during the supplanting of aboriginal populations in the Americas and Australia, for example.

To Arms!

Even if we concede – without any real scientific justification – that Neanderthal populations of this time were genetically less dynamic – why not? – we would still have to explain why and how this humanity continued to survive in Spain and Siberia, and beyond, without encountering major problems, for hundreds of thousands of years. Even here their eventual demographic decline can also be attributed to the particular abilities of *sapiens* societies, who seem to have possessed clearly superior hunting technologies, allowing them much greater and easier access to substantial animal resources. We must assume that this access to resources could have significantly contributed to the demographic expansion of *sapiens* groups.

The question of weapons seems to me fundamental to understanding not only the phenomena of replacement but also the functioning of Neanderthal or *sapiens* societies as a whole. Hunting a horse or bison using technology requiring a direct physical contact with the prey is one thing. Effortlessly killing whole series of large herbivores in half a day with weapons employing high-energy mechanical propulsion – bow or spear – is quite another. The technologies in themselves were enough to assure an abundant, efficient and well-planned access to animal resources, as suggested by Laure Metz in her doctoral thesis 'Neanderthal Armed?'. Such easy and planned access to resources must have had radical repercussions in terms of the demographic divergence of these populations.

We can pose this hypothesis because it seems that the Neanderthals had remarkably few weapons. Take any series of tens of thousands of flint objects from any Neanderthal collection and you will find only the very occasional weapon. I would even go so far as to say that you will find them only if you really look hard for them. As objects they are consistently heavy, oddly shaped and often lacking in any technical refinement. At best, assuming they are indeed weapons, we are talking about ends of lances that would be used for stabbing rather than throwing. This has been discussed in a number of recent studies, which have come to somewhat binary conclusions. The frantic quest for Neanderthal weapons is rather like the quest for Neanderthal art: a tenuous project, as we discussed earlier. A lot of the scientific literature draws the same parallel between art and weapons. We might sum it up like this: if the Neanderthals possessed weapons, they were like us, and so we can conclude unequivocally that this population was in no way different from contemporary *sapiens* populations.

Such conclusions and lines of argument seem remarkably paradoxical. If we step back from the scientific controversies, we very quickly see that the empirical evidence in fact points in the opposite direction. Despite a systematic search for weapons in some very well-endowed archaeological ensembles, the fact that such objects are so rare suggests that weapon production was a very marginal activity for the Neanderthals. The question of Neanderthal weapons remains little understood to this day. The picture we get is one of rudimentary technologies based on the production of massive lances or javelins that required a close contact with the game being hunted. The hunt would have been carried out using a lance and would have involved the hunters approaching their prey and taking them on at close quarters. In Mandrin as elsewhere, when we discover arms in the Neanderthal layers, they are always massive objects and would have been used as pole weapons. As access to meat resources was probably limited, Neanderthal groups would have remained modest in size. When we look for *sapiens* arms, just a cursory glance at any collection will reveal a large number of objects that could potentially be used for hunting purposes. In Neanderthal collections, you need to dissect huge corpora of flint pieces to uncover the odd vaguely diagnostic trace of weapon use. The rarity of arms in the Neanderthal archive is striking.

Weapons are central to every recent European Palaeolithic society, from their origin with the Proto-Aurignacian more than 42,000 years ago and probably more than ten millennia earlier, with the societies we might label Pre-Proto-Aurignacian. With this in mind, we can link the beginning of the recent Palaeolithic with the development of mechanical propulsion, which intrinsically involved standardization in these crafts. The modern humans who colonized Europe produced weapons that were both more numerous and more effective, and it was this that seems to have brought about a major shift in the technological and probably also social organization of their societies. Mechanical propulsion, and more specifically archery, allows technological progress towards microlithization, serial production and standardization. System change is here directly related to the ballistic constraints induced by the very high-energy propulsion technologies.

Such realities, such technical and ballistic constraints, reconfigured not just the technical traditions of *sapiens* societies, but also their relations with the animal world and their ability to plan for their needs. Technology had an impact not just on artisan production but also on all the logistical and social relations that the populations were able to form among themselves and with their natural environment. From this we can get an idea of the structural elements underpinning the upheavals in these societies, in their organization, their values and their rational ability to control the natural world in which they developed. This question of weapons leads on to standardization, access to proteins, planning and reproductive and demographic success.

So we can say that there was a fundamental, and potentially structural, divergence here between the Neanderthal and *sapiens* societies. The Neanderthals evidently did not hunt in the same way as the *Sapiens* and probably had a different relationship with their game. My point is that they had fundamentally distinct modes of acquisition at the time when they encountered each other in Europe. What happened here is like what happened in the colonization of the Americas, when societies armed with bows encountered Europeans equipped with guns. The power relationship was highly skewed and asymmetric. Weapons technology also played a fundamental part in the success of this colonization not only during conflicts but also when, in the nineteenth century, the colonists decided to hunt the bison to extinction, which inflicted famine on the Indians. As Colonel Dedge put it in 1867: 'Every buffalo dead is an Indian gone.' The railways that were built across the United States from east to west in fact followed the trails made across the Plains by herds of bison, generation after generation. As part of the conquest of the West, bounties were paid on each head of bison, with the result that populations collapsed from tens of millions to just a few hundred within a century, forcing the Indian nations to sign inequitable annexation treaties which were almost never respected. To a different degree, of course, the territorial expansions of *sapiens* populations were by definition markers of extremely dynamic societies which possessed weapons technology that was unknown to, or simply rejected by, the native groups. Such expansions and dynamism are not positive features of these so-called modern societies but in fact have a terrible impact on aboriginal populations.

The really interesting question is not to do with the weapons themselves but what they tell us about the behaviour of these societies and these populations. If we assume, as I do, that weapons technology provides a major key to understanding how these human replacements happened, we are led to conclude that the ability to access animal resources was potentially more limited in the local Neanderthal groups, which further implies that these groups were potentially numerically much reduced. By the time of their encounter there was an imbalance between the two populations, both numerically and technologically. That is also reminiscent of the colonization of the Americas, but the differences then were exclusively technological and cultural, which is not the case concerning the encounter between Neanderthal and *Sapiens*.

The Fundamental Structures of the Two Humanities

We should add another layer to the discussion here: that of the biological diversity of the two humanities. In addition to these divergences, did the Neanderthals have a certain way of being in the world, of fitting into and understanding their environment?

In other words, was there a Neanderthal ethology that was different from that of *Sapiens*? The fact that no Neanderthal population ever produced weapons in a systematic way, that they were never interested in a normalized, standardized form of production that would have enabled them to acquire technical systems specifically geared to the acquisition of their prey, leads us to suppose that Neanderthal populations had structures of understanding the world that were unique and distinct from those of our *sapiens* ancestors. Beyond technological systems, beyond knowhow and traditions, beyond modes of acquiring protein, there is something much deeper at work than a simple cultural phenomenon: we can actually perceive *another* humanity. Neanderthal societies organized themselves as they went along, they 'went with the flow', showing only a passing interest in the modes of planning that are still a salient feature of our current societies. Observation of *sapiens* technological systems reveals systematic modes of planning and standardization. *Sapiens* artisans were technically remarkable. It is clear in their objects: their artefacts, their artistic forms, still instinctively resonate with us. But their work also seems surprisingly tiresome and depressing to the specialist in Neanderthal societies. To put it bluntly, the *sapiens* crafts of the Palaeolithic are all about us. They speak only of us, of our societies, of our ways of being. If we look at a hundred flints, we can easily grasp the techniques involved, and we know the next 100,000 will be exactly the same. We understand instinctively what the maker was trying to do, which is never the case with Neanderthal artefacts.

We are very close here to a fundamental divergence between our humanities. It is disturbing to think that barely a few tens of thousands of years ago other complete humanities, with their own cultures, traditions and artefacts, were not human in the narrow and banal sense we are familiar with now. We should be able to represent these other humanities without immediately foisting our projections, our subjective hierarchies, our notions of superiority and inferiority on to them. The human–non-human binary opposition becomes evidently false the further back in time we go. Indeed, as we regress through time, we reach a moment, a moment not too long ago, when there is no recognizable human or humanoid life-form on the planet. There is evolution – technical, social and biological – and we can't simply project our schemas, our definitions, on to these distant humanities. Even less so on to biologically fossil humanities. This archaeology teaches us that our classifications are just rigid pigeonholes that do nothing except order a history whose immense complexity escapes us fundamentally.

Yet there are many distinct Neanderthal cultures. We *Homo sapiens citadensis* are absolutely normalized by our society. You just have to walk down any street to see that our expression of diversity is more or less limited to the case we put on our smartphones or the colour of car we choose. Our societies do not tolerate any real expression of plurality. In western society women can now wear their hair long or short, trousers or skirt, make-up or no make-up, but the same is more problematic for men. We belong to an over-normative society, and that is basically true of all *sapiens* societies, at all times: difference is frowned upon and only superficially tolerated at the periphery. This is probably a matter of ethology, a phenomenon deeply embedded in our genes and not simply a cultural feature. We live lives restricted by our norms. We all confine ourselves to specific categories.

With the Neanderthals, this seems to me altogether more

subtle. I don't mean that they just picked up any old bit of limestone and gave it a rough spit and polish. They had crafts that required great technical skills; they were great carvers. They were excellent artisans, and some of their artefacts are technically challenging to reproduce today. They had remarkable knowhow, and this was passed on. These artisans produced whole categories of objects that would be used every day in a series of activities (cutting meat, tanning hides), but without ever carrying out these activities a second time with an identical tool. No two Mousterian tools are alike, and that is remarkable. And yet we find very expressive styles which could be considered to be characteristic of different human groups. There is no doubt that particular skills were passed on. But this was a culture without normalization, without standardization, without systematic repetition, without that quasi-industrial character that defines both prehistoric *sapiens* cultures and modern-day societies.

We have before us the fundamental structures of these humanities: the profound differences between Neanderthals and *Sapiens*. Each Neanderthal tool is a creation in itself. It plays with the natural forms of the raw material, with the texture of the rock, with its colours, with its touch. There is a balance, an absolute perfection to the Mousterian object, almost indefinable but nonetheless present, which reveals a remarkable way of seeing the world. The constant play that these people established between the materials they used and their technological traditions brings us face to face with a creativity that is beyond us. And this infinite playful production of original works, which is nonetheless based on well-defined traditions, enters into a dialectic with the materials, the textures, the colours of the rocks, which guide, or participate in, the balance of the whole creation.

We find ourselves face to face with an infinite creativity, beyond compare with the technical products of our societies. The subtle dialectics between the object and its material shows clear variations which are partly to do with the inherited traditions of these groups. This is also, I think, what makes each Neanderthal object distinctive and adds a systematic fluidity to the artisan's technical project. Each tool we have found is de facto a unique object. The singular properties of these crafts are perceptible in all the cultural traditions of Neanderthal groups, even if they are to do with spheres of daily activity that are fundamentally repetitive. The object is simultaneously over-invested technically and unique in its construction. So in that sense it is necessarily related to the mental structures of the artisans themselves, whatever technical traditions they might have inherited. There is an absolute artisanal freedom, and probably a very rich freedom of thought about the world. I would suggest that the artisanal production of objects by the Neanderthals reveals a perception of reality that has no structural echo in what we see in *sapiens* societies, whether Palaeolithic or modern-day. In this the Mousterian object is most like certain Eastern traditions such as the Japanese *shibui* and *ma* or the Maori *mana*, which give us a preliminary idea of what Neanderthal material production was like. If it is possible to propose parallels between Neanderthal craft constructions and certain modern-day spiritual currents, should we then consider that there is no structural distinction in ways of being between the world of *Sapiens* and that of the Neanderthals? I don't think that this interpretation is correct. We are weighing up all Neanderthal societies, in which we have documented a wide range of traditions, against some very limited currents of human thought. We are comparing a Neanderthal structure with some isolated and marginal expressions in modern-day culture, a structure that was common to the whole of a biological population with certain very circumscribed cultural sensibilities. The quantitative approaches classically adopted in analyses of these fossil societies are then not able to take account of such ethological peculiarities, any more than they could take account of *shibui* or *mana* from one or from a thousand objects.

This raises a question about what can't be quantified: these currents, these sensibilities, these concepts, which exist in their own right and from which the very framework of whole societies can be articulated. It is not the tangible reality of these concepts that is in play here, but the ability of would-be scientific quantitative methods to take account of the structural elements of human societies. However, neither the approach based on sensibilities nor that based on aesthetics would enable us to map the mental universe of Neanderthal populations. Aesthetics concentrates on the most superficial texture of the Mousterian object whereas an analysis of the mental structures at work suggests that there should be a whole programme of research in human ethology, a whole discipline in itself. And yet, the very analysis of technical systems, which is powerful even though it has been emerging for forty years, has not provided the means to conceptualize the possibility of ethological divergence. It is not that this tool is unable to do so, just that it hasn't been used to this end. It has never been used in a way that is not haunted by our mental schemata, by the projection of our limited rationality. It is this same false rationality that until recently led us to read the economy of traditional societies as a subsistence economy. It wasn't until 1972, and the remarkable study by Marshall Sahlins, *Stone Age Economics*, that we started to understand how we project our western assumptions on to societies that are essentially unknown to us.

Although we have made some progress in our understanding of traditional societies, our understanding of fossil societies appears to still be narrowly confined by our western conceptions. If we do not make ourselves ask certain (im)pertinent questions, then our analysis of distant Neanderthal crafts cannot grapple with these mental worlds.

The intelligence of the Neanderthal hand, which we can note even if we can't define it, goes well beyond a purely technical intelligence and transcends notions of aesthetics, of balance, of the irrational function of the sign, showing the remarkable character of the manufactured object and that of the person who created it. The artisanal and artistic creations of *Sapiens* are beautiful, but they are beautiful and nothing more . . . They rarely go any further. For *Sapiens*, art is just an expression and affirmation of ego. It seems to me that Neanderthal creativity and sensibility greatly transcend the egocentric products of our societies to achieve a form of universal beauty, in which the ego is no longer central but is given a more peripheral position. According to this logic, art, symbolism and the sign are simply not separable from everyday artisanal creations. They don't need to be. They arise from the same function. Technical and artistic expressions are integrated in the same global logic. Art for art's sake tells us about the artists. Neanderthal art, art fused with technology, does not speak about the person, the individual, the ego, but exclusively about the ways of being in the world of the group as a whole.

If these conclusions are correct, then we have touched upon a rather unexpected definition of our own species and a characterization of a very different humanity for which we do not need to look for the art and the symbol as narrowly, restrictively defined by our societies. The question of Neanderthal art here perfectly echoes the definition of the Neanderthal weapon that we have just proposed. If there is a Neanderthal art, it seems, as with their weapons, limited to very little: remarkable objects that have been collected, scrapings, indefinable marks, where the meaning is always debatable. Singular objects, shells, claws, minerals, whose objective purpose, intention, is never expressed by the craftsman, never proven objectively. We are still waiting for the first Neanderthal hole to hang a tooth, a shell, a bone. And we line up these handfuls of vague, indefinable marks and, like their weapons, make them say: 'Look, they are like us.' And the more we line up these rare, randomly shaped weapons and these paradoxical, unconvincing art objects, the more we see these people breaking out from our narrow interpretative frames, from our simplifying pigeonholes.

No, the Neanderthals are not ersatz *Sapiens*. Not only are they different, but in many mental aspects they overshadow *Sapiens* – in their total, ongoing creativity, essentially free from the ego which structures so much of the differentiation of group and individual in *Sapiens* populations. In this sense, and comparatively, our population is very superficially creative. Indeed, it can be argued here that, in the fields of creativity, *Sapiens* was probably no match for Neanderthal populations and was in all likelihood intellectually inferior. But this wasn't the case for the material rationalization of the world, in which, perhaps, the Neanderthals came off second best.

It seems that we *Sapiens* are not very good at embracing difference, which in itself makes any representation of another humanity and another ethology very difficult. It feels self-destructive to imagine any real divergence. But as we go back in time we can't conceive of past humanities as one monolithic block, with no variations. This is our creationism 2.0. Palaeo-genetics doesn't tell us that a Neanderthal is a *Homo sapiens*, but that the two populations diverged over the course of hundreds of millennia, each evolving differently, in parallel, adapting in their own ways to dissimilar environments.

These biological divergences of the populations described by genetics are echoed in all the spheres that organize human societies: in artefacts and weapons, but also in art. Each time that teams of scientists have tried to find signs of resemblance which would bring the Neanderthals closer to us, they have only come up with variations on certain simplified themes which in my opinion do not justify us thinking about Neanderthals as another version of us. And that is fortunate when you think about it. A few lines marked on a piece of bone or a rock don't indicate artistic conception in the narrow sense we understand it today. Between our last Neanderthals in the Mandrin cave and the artists of the Chauvet cave there are *only* 6,000 years, in other words, nothing at all. These last Neanderthals, however, had the same ways of inhabiting their world as populations from 100,000, 200,000 or 300,000 years ago. And here, if we don't make the effort to think about these societies in terms of their material production as a whole and just focus on a few insignificant marks on bones, if we don't try to consider this material production for what it is, then we end up skating over the subject.

It is not a question of knowing how Neanderthal artefacts resemble our own, but of defining how they are fundamentally structured. Focusing on a handful of anecdotes is like analysing the leaf of a tree without analysing the forest of which it is a part. The Neanderthals have left millions of remarkable artisanal objects which allow us to come face to face with the fundamental structures of thought and the true nature of this extinct humanity.

Epilogue: Liberating the Creature

I have been digging in Neanderthal sites for nearly thirty years, two to four months a year, and I don't buy into the prevalent idea of a clean-cut, respectable Neanderthal, one limited to us. The usual argument for this standard image is that it is a way of rehabilitating the Neanderthal, who was once defined as different, as little more than a beast, a view now seen as a product of the racist discourses of the nineteenth and twentieth centuries. But in rushing to challenge these views we avoid analysing what racism really is.

And we refuse to look closely at the profound process of projection taking place in front of our eyes. Racism is the refusal of difference. The rejection of difference, its expulsion from humanity. Racism is those old images of Plains Indians trussed up in three-piece suits: just like us.

By making the Neanderthal into another us we reveal this unconscious racism that still fundamentally structures the schemas of our own society, incapable as we are of imagining any form of alterity. There can't be human difference because we limit our definition of humanity to the narrowest frame of reference.

So here are the two conceptions, in direct opposition. The one that I express in these pages is not simply a way of seeing subjective realities. It is objective in the sense that it is the fruit of a train of unprejudiced thought which has ripened very gradually over decades of daily contact with Neanderthal remains. It seems to me now that there are several gradations of ways of being in the world. There is no objective, logical, rational reason why these populations, which evolved independently for hundreds of millennia, should have developed ways of being in the world which are the same as ours. Here again, our unconscious assumptions lead to a persistent unconscious creationism in our society: all intelligent creatures must naturally tend to become, and then limit themselves strictly to, what it is that we have become.

New perspectives are emerging. Could the Neanderthals be finally about to be freed from the sad mime artist role which we have imposed upon them? Will they finally achieve total freedom?

I hope so. But for now, I fear the creature will remain a prisoner of our prejudices for a long time yet. You can't get free of yourself so easily.

Further Reading

All the chapters in this book draw upon a large documentary corpus, which can be accessed in specialist English-language periodicals. The book expresses my personal views on these distant populations and evokes some of the routes I have followed during my time as a researcher. So this work is not inspired by any previous publication on the profound structure of Neanderthal societies and doesn't exhaustively address all areas of research, whether mainstream or alternative.

The references listed below refer to theories or to human experiences which offer ways to touch, albeit ever so slightly, upon certain deep structures of our own way of being in the world which have been tackled here.

Albert, B., Dreyfus-Gamelon, S., Razon, J.-P. (eds.), 'Chroniques d'une conquête', *Ethnies*, 1993, 7 (14), pp. 1–118.

Artières, P., *Le Dossier sauvage*, Gallimard, 2019.

Bachelard, G., *La Formation de l'esprit scientifique. Contribution à une psychanalyse de la connaissance*, Vrin, 2004.

Beuys, J., Harlan, V., *Qu'est-ce que l'art?*, L'Arche, 2011.

Catarini, S., *Les Non-Dits en anthropologie, suivi de Dialogue avec Maurice Godelier*, Thierry Marchaisse, 2012.

Descola, P., *Par-delà nature et culture*, Gallimard, 2015.

—, Taylor, A. C. (eds.), *La Remontée de l'Amazone. Anthropologie et histoire des sociétés amazoniennes*, special number of *L'Homme*, 126–8, April–December 1993.

Godelier, M., *La Production des Grands Hommes*, Flammarion, 2009.

—, *Les Tribus dans l'histoire et face aux États*, CNRS Éditions, 2010.

Hell, B., *Le Sang noir. Chasse et mythe du sauvage en Europe*, Flammarion, 1994.

Kroeber, T., *Ishi. Testament du dernier Indien sauvage de l'Amérique du Nord*, Plon, 1968.

Lee, R. B., DeVore, I. (eds.), *Man the Hunter*, Aldine Publishing Company, 1968.

Lévi-Strauss, C., *The Elementary Structures of Kinship* (1949), Beacon Press, 1971.

—, *Race and History* (1952), Franklin Classics, 2018.

—, *Tristes tropiques* (1955), Penguin, 2011.

—, *Structural Anthropology* (1958), Basic Books, 1974.

—, *Wild Thought* (1962), University of Chicago Press, 2021.

— (ed.), *L'Identité. Séminaire interdisciplinaire dirigé par Claude Lévi-Strauss, professeur au Collège de France. 1974–1975*, Grasset, 1977.

—, *The View from Afar* (1983), Basic Books, 1985.

Lewis, M., Clark, W., *The Essential Lewis and Clark*, ed. Anthony Brandt, National Geographic, 2018.

Loeb, A., *Extraterrestrial: The First Sign of Intelligent Life Beyond Earth*, John Murray, 2022.

Malaurie, J., *Les Derniers Rois de Thulé*, Plon, 1976.

—, *L'Appel du Nord. Une ethnophotographie des Inuits du Groenland à la Sibérie : 1950–2000*, La Martinière, 2001.

— (ed.), *De la vérité en ethnologie . . . Séminaire de Jean Malaurie 2000–2001*, Centre d'études arctiques / EHESS / Economica, 2002.

Morris, D., *The Biology of Art: A Study of the Picture-making Behaviour of the Great Apes and Its Relationship to Human Art*, Methuen, 1961.

—, *The Naked Ape: A Zoologist's Study of the Human Animal* (1968), Delta, 1999.

Musée de l'Homme (exhibition catalogue), *Arts primitifs dans les ateliers d'artistes*, Société des amis du musée de l'Homme, exhibition curator Marcel Evrard, 1967.

Plumet, P., *Peuples du Grand Nord*, Errance, 2004, 2 vols.

Quppersimaan, G., *Mon passé eskimo*, Gallimard, 1992.

Sahlins, M., *Stone Age Economics*, Routledge, 2017.

Solecki, R. S., *Shanidar, the First Flower People*, Alfred A. Knopf, 1971.

Wachtel, N., *The Vision of the Vanquished: The Spanish Conquest of Peru through Indian Eyes, 1530–1570*, Harvester Press, 1971.

Acknowledgements

As a Neanderthal hunter, I am constantly engaged in my research in caves, my analyses of millions of knapped flints, my reports, my scientific writings, my evaluations of other scientific works, and these activities have thus far kept me from sharing my knowledge with a wider audience.

It was once again, of course, unexpected encounters that motivated me to share my thoughts. Among these many encounters, that with the Rosen family was a true catalyst, and I express my profound gratitude to Élie, Michel and Laurence. I am also very grateful to David Watson and Casiana Ionita for their work on the English edition.